Combinatorial
Algorithms

Combinatorial Algorithms

Luděk Kučera

Faculty of Mathematics and Physics,
Charles University, Prague,
Czechoslovakia

Adam Hilger, Bristol and Philadelphia

British Library Cataloguing in Publication Data

Kučera, Luděk
Combinatorial algorithms.
1. Graphs. Numerical solution. Algorithms
I. Title II. Kombinatorické algoritmy
English
511'.5

ISBN 0-85274-298-3

Library of Congress Cataloging-in-Publication Data

Kučera, Luděk.
[Kombinatorické algoritmy. English]
Combinatorial algorithms/Luděk Kučera.
Translation of: Kombinatorické algoritmy.
Bibliography:
Includes index.
ISBN 0-85274-298-3
1. Electronic digital computers–Programming. 2. Algorithms.
3. Graph theory. I. Title.
QA76.6.K8413 1989
005.1'2–dc19   88–19307 CIP

Published under the Adam Hilger imprint by IOP Publishing Ltd
Techno House, Redcliffe Way, Bristol BS1 6NX, England
335 East 45th Street, New York, NY 10017-3483, USA
Published in co-edition with SNTL, Publishers of Technical Literature,
Spálená 51, 113 02 Prague, Czechoslovakia

Printed in Czechoslovakia

Contents

Preface

This book has evolved from lectures held by the author at the Faculty of Mathematics and Physics of Charles University, Prague. Although it is directed primarily to students of computer science, it will also be useful to programmers and other workers in the area of computers. When writing programs, we often encounter problems which have the following property: they can be formulated in the language of graph theory, but it is not easy to find sufficiently effective methods for their solution. Let us mention, for example, transport problems, analysis of networks, problems of time schedules (whether in economy or in operation systems of computers), etc.

This book is devoted to the solution of problems presented by the theory of graphs. It can be considered as a continuation of current textbooks on graph theory; see, for example, references [3, 4, 7, 8, 12, 18, 28, 30 and 41].

The basis of the book is formed by Chapters 6, 7 and 9. In Chapter 6 we describe problems for which a good algorithm is known: searching for the shortest and critical paths in a graph, flows in networks, searching for connected components of a graph (which has direct applications in reliability theory and, besides, is important for a decomposition of complex problems into simpler ones) and processing planar graphs (important, for example, for the automatic design of printed wiring and integrated circuits).

However, for the majority of problems in graph theory no algorithm is known today which would find exact, or optimum, solutions and would be sufficiently fast to process graphs of hundreds and thousands of vertices. In Chapter 7 we shall explain the reasons for this situation. We will describe the class of so-called NP-complete problems, including such classical problems as the travelling salesman problem, the general problem of integer programming, and a number of its special cases, the search for a hamiltonian circuit in a graph, the determination of the chromatic number of a graph, a number of problems concerning time schedules, etc. All those problems are, in a way, of the same degree of difficulty, and it seems that their computational complexity is so large that it is beyond our contemporary possibilities of computation (although this pessimistic statement has not yet been proved − nor disproved). Chapter 8 is devoted to a deeper analysis of the questions studied in Chapter 7. Both these chapters represent the most theoretical part of our book.

In contrast, Chapter 9 is oriented to practical problems. Here, we introduce heuristic and approximate methods for the solution of some of the NP-complete problems and the problem of searching for an isomorphism of two graphs. These methods are usually very fast and make it possible to process even large graphs, but the solution obtained is usually not the best one possible. Chapter 10 then studies the probabilistic analysis of some algorithms presented in Chapter 9.

Chapters 1 and 2 present a concise introduction to the theory of graphs and models of computation. In Chapters 3, 4 and 5, which are more specialised, we introduce important techniques necessary for designing algorithms for graph processing. The first of these is devoted to the problem of an operative storing of data. Some of the data structures described there are themselves based on graphs (usually directed graphs) and the algorithms of manipulation with the data structure represent a nice sample of combinatorial algorithms. Chapter 4 investigates methods of searching a graph and Chapter 5 is devoted to methods of fast sorting of numbers according to their size.

It is necessary to mention some of the problems which have not been included in the book, although they belong to the field of problems covered by it. This concerns, in the first place, the problem of linear programming since a number of problems which we are going to study are just special cases of it. The problem of linear programming has a satisfactory solution, e.g. by the simplex algorithm (verified by practical experience) or by the theoretically important algorithms of Chatchiian or Karmakar. On the other hand, we will show in this book that, for problems like that of the travelling salesman or maximum flow in a network, there are algorithms which are faster than those obtained by applying the general procedures of linear programming.

More space could have been assigned to the problem of the average behaviour of algorithms. This, however, would require an introduction to probability theory, and for this reason Chapter 10 has rather the style of a survey article.

The evaluation of algorithms in this book is almost exclusively based on their speed, and we do not derive results concerning the requirements on memory space (which usually is a less important limiting factor).

We have also omitted the interesting field of problems concerning the relationship of requirements on the running time and the memory space, and the design of algorithms requiring the minimum space.

The development of chips and micro-processor systems brings with it the possibility of accelerating the computation by means of parallelism. Another field of problems which has arisen in connection with the development of micro-electronics is the design of circuits of VLSI (very large scale integration) where the dominant question is the relation of the speed of algorithms realised by the circuit and the surface of the circuit. By enlarging either the surface or the scale of integration, we can speed up the computation using a larger number of memory elements or processors or filters, but that surface can also be wasted by an increase in the number of interconnections of subsystems should the design be unsuitable or too complicated. These problems have not been included in this book because the theory is too extensive.

The field of problems to which the book is devoted has recently been growing dramatically. Most results can be found only in specialised journals, technical reports and proceedings of conferences. Thus one feels there is a lack of textbooks and monographs dealing with the question of computation complexity of problems of graph theory in a systematic manner. The main goal in writing this book was to try to remedy the situation somewhat.

Luděk Kučera
Prague

1

Basic concepts and results

The first chapter is devoted to a survey of concepts and results which will be needed later but which do not belong to the actual field of problems investigated in this book. The material covered by the present chapter is probably known to the reader and, if not, he can find more detailed information in the textbooks listed at the end of the book. We shall also mention the origin of some of the problems studied later.

1.1 Some basic concepts

In this section we mention briefly some of the mathematical concepts and notations which will be used throughout the book.

We assume that the reader is familiar with the set-theoretical concepts which now belong to the curriculum of elementary schools, for example, with the following:

set
subset (denoted by $X \subset Y$),
proper subset ($X \subsetneqq Y$, i.e. $X \subset Y$ and $X \neq Y$),
intersection of sets ($X \cap Y$),
union of sets ($X \cup Y$),
difference set ($X - Y$).

The symbol $\emptyset$ denotes the empty set. If x is an element of a set X, we write $x \in X$. If $X \cap Y = \emptyset$, we say that the sets X and Y are *disjoint*.

The set of all elements x of a set X which have a specified property V is denoted by $\{x \in X \mid x$ has the property $V\}$. If the set X is obvious, we also write $\{x \mid x$ has the property $V\}$.

Sets can be finite, or infinite. We restrict the collection of infinite sets which will be used below to a few, for example, the set of all natural numbers, or integers, or real numbers, and formal languages as introduced in Chapter 2, and some other sets. Otherwise we shall be working solely with finite sets, e.g. whenever we construct a graph, or a similar object.

The *cartesian product* of sets X and Y, denoted by $X \times Y$, is the set of all ordered pairs (x, y) where $x \in X$ and $y \in Y$. The cartesian product of three or more sets is, analogously, defined as the set of all ordered triples (in general, ordered n-tuples) of elements of the corresponding sets.

An arbitrary subset R of the cartesian product $X \times Y$ is called a (*binary*) *relation* between the sets X and Y. If $X = Y$, then R is called a relation on the set X. Various conditions are imposed on relations on a given set X, among the most important of which are the following:

reflexivity: for each $x \in X$, $(x, x) \in R$;
transitivity: if $(x, y) \in R$ and $(y, z) \in R$, then $(x, z) \in R$;
symmetry: if $(x, y) \in R$, then $(y, x) \in R$.

A relation which is reflexive, transitive and symmetric is called an *equivalence*. Such a relation uniquely determines a decomposition of the set X into pairwise disjoint classes of the following form:

$$\bar{x} = \{y \in X | (x, y) \in R\},$$

where x ranges over the elements of the set X. In fact, it is easy to verify that for arbitrary elements $x_1, x_2 \in X$ either $\bar{x}_1 = \bar{x}_2$, or the sets $\bar{x}_1$ and $\bar{x}_2$ are disjoint. Conversely, each disjoint decomposition of the set X determines an equivalence relation on X: two elements are equivalent if they belong to the same class of the decomposition.

A *mapping* f from a set X into a set Y (denoted by $f: X \to Y$) is a rule which to each element x of X assigns a unique element y of Y, denoted by $f(x)$. A mapping f can be considered as the relation between the sets X and Y consisting of all the pairs $(x, f(x))$, where $x \in X$. In the case where for each $y \in Y$ there exists at least one $x \in X$ such that $f(x) = y$, we say that the mapping f is onto Y, or, a *surjective* mapping. In the case where for two arbitrary distinct elements x_1 and x_2 of the set X we have $f(x_1) \neq f(x_2)$, we say that f is one-to-one, or an *injective* mapping. An injective mapping of a set X onto a set Y is called a *bijection* or a bijective mapping.

If r is a real number, then $\lceil r \rceil$ will denote the least integer larger than or equal to r, and $\lfloor r \rfloor$ the largest integer smaller than or equal to r. If n is a natural number, i.e. $n = 1, 2, 3,...$, then $n!$ (the factorial) denotes the product $1 \cdot 2 \cdot 3 \cdot ... \cdot n$. By a *polynomial* is meant a function assigning to each real number x the value

$$P(x) = a_n x^n + a_{n-1} x^{n-1} + ... + a_1 x + a_0$$

where $a_0, a_1, \ldots, a_n$ are numbers (real numbers, in general). If $a_n \neq 0$ then n is called the *degree* of the polynomial.

Although this is somewhat unusual, the symbol log will denote the binary logarithm, i.e. $\log x = \log_2 x$ is a positive number y with $2^y = x$. The natural logarithm, i.e. the logarithm with base e, will be denoted by ln, and the decadic logarithm will not be used in our book. In this book, exp x means 2^x (contrary to usual notation exp $x = $ e^x). This notation will be often used if x is a complicated expression.

If f and g are functions assigning real values to natural numbers, then we use the following notation:

$f = O(g)$ if there exist natural numbers n_0 and K such that for each $n \geq n_0$ we have $f(n) \leq K g(n)$,

and

$f = \Omega(g)$ if there exists a natural number n_0 and a real number $k > 0$ such that for each $n \geq n_0$ we have $f(n) \geq k g(n)$.

The first statement means that, up to a multiplicative constant, the function f does not grow asymptotically more quickly than the function g, and the latter means that f does not grow more slowly. We further write

$$f = \Theta(g) \text{ if both } f = O(g) \text{ and } f = \Omega(g),$$

and this means that f grows asymptotically as quickly as g.

In some parts of the book we are going to use formulae of predicate calculus. Each formula depends on logical variables which will be denoted by lower-case letters $x, x_1, x_2, \ldots, y, z, u, \ldots$. A logical variable takes two values, true and false. The variables and the logical constants true and false constitute the so-called atomic formulae. Formulae are formed from atomic formulae by repeated use of the following operations (on formulae f and g):

$\neg f$,   negation (f does not hold),
$f \vee g$, disjunction (f or g),
$f \wedge g$, conjunction (f and g),
$f \rightarrow g$, implication (if f, then g),
$f \equiv g$, equivalence (f if, and only if, g).

The negation of f depends on the same variables as f, the other formulae above depend on those variables on which either f or g depend.

If truth values are assigned to all variables, then the truth value of each atomic formula is the corresponding value of the variable or the constant. The truth value of each formula is then specified by the following rule:

operation has value 'true' if, and only if,

$\neg f$	f has value 'false'
$f \vee g$	f, or g, or both have value 'true'
$f \wedge g$	f and g both have value 'true'
$f \rightarrow g$	either f has value 'false', or g has value 'true'
$f \equiv g$	f and g have the same value.

As an illustration of the construction of a more intricate formula (in variables x, y, z, u and v), let us consider the following scheme:

$$x \quad \neg y \quad z \qquad\qquad x \quad u \quad v$$
$$x \quad \neg y \wedge z \qquad\qquad x \quad u \wedge v$$
$$x \rightarrow (\neg y \wedge z) \qquad\qquad x \vee (u \wedge v)$$
$$(x \rightarrow (\neg y \wedge z)) \equiv (x \vee (u \wedge v)).$$

Assuming, for example, that the variables x, y and v have value 'true' and the variables z and u have value 'false', then we obtain the following table of truth values:

$\neg y$	false
$\neg y \wedge z$	false
$u \wedge v$	false
$x \rightarrow (\neg y \wedge z)$	false
$x \vee (u \wedge v)$	true

and hence, the whole formula has value 'false'.

If a formula has, for given truth values of the variables, the value 'true', then we say that it is *satisfied*; if its value is 'false', then it is *unsatisfied*. A formula satisfied for arbitrary truth values of the variables is called a *tautology*. The following formula is an example of a tautology:

$$(x \rightarrow (y \rightarrow z)) \equiv ((x \wedge y) \rightarrow z).$$

If there exist truth values of the variables such that the truth value of a given problem is 'true', then the formula is said to be *satisfiable*. The problem of determining whether a given formula is satisfiable, or a tautology, will play an important part in Chapter 7.

Atomic formulae and their negations will be called *literals*. Literals and disjunctions of two or more literals are called *clauses*. A formula which is equal to a clause or a conjunction of two or more clauses is said to be in *conjunctive normal form*. The formula above is not in conjunctive normal form. An example of a formula in the conjunctive normal form is the following:

$$(x \lor \neg y \lor z \lor u) \land (\neg x \lor \neg u \lor v) \land (x \lor y \lor \neg v) \land (\neg y \lor \neg u).$$

A *quantified* formula is an expression derived from atomic formulae by the above operations of negation, disjunction, conjunction, implication and equivalence, together with the following two operations:

$\forall x\, f$, the *general quantification* (f is true for each x)

and

$\exists x\, f$, the *existential quantification* (f is true for some x).

All the instances of the variable x in the formula $\forall x f$ and $\exists x f$ are said to be *bounded*; the other variables occurring in a formula are said to be *free*. A quantified formula is closed if all variables occurring in it are bounded. The formulae $\forall x f$ and $\exists x f$ depend on all variables on which f depends, except the variable x.

To avoid complications we shall describe how to determine the truth value of a quantified formula only when it has the following form:

$$F = Q_1 x_1 \ldots Q_n x_n\, f(x_1, \ldots, x_n, y_1, \ldots, y_m), \qquad n \geq 1.$$

Here $Q_1, \ldots, Q_n$ are the quantifiers ($\forall$ or $\exists$) and f is the quantifier-free formula depending on the variables $x_1, \ldots, x_n, y_1, \ldots, y_m$. (The formula F depends on the variables $y_1, \ldots, y_m$.) Put

$$F' = Q_2 x_2 \ldots Q_n x_n\, f,$$

then F' depends on the variables $x_1, y_1, \ldots, y_m$.

We say that the formula F has, for given truth values of the variables y_i, the truth value 'true' if, and only if,

either $Q_1 = \forall$ and F' is satisfied both for the given truth values of y_i and $x_1 =$ true, and the given truth values of y_i and $x_1 =$ false,

or, $Q_1 = \exists$ and F' is satisfied either for the given truth values of y_i and $x_1 =$ true, or for the given truth values of y_i and $x_1 =$ false.

A quantified formula in conjunctive normal form is a formula of the form

$$\exists x_1 \forall x_2 \ldots \exists x_{2n-1} \forall x_{2n}\, f(x_1, \ldots, x_{2n}),$$

where f is a quantifier-free formula in the conjunctive normal form, depending on the variables $x_1, \ldots, x_{2n}$. Such a formula is closed, and in Chapter 8 we shall study the problem of whether it is true, i.e. has value 'true'.

1.2 Graph theory

For the formulation of most of the problems which we are going to study we shall use the language of graph theory and we shall work with properties of graphs, sometimes simple and sometimes more complex. The reader who finds the following presentation too brief can consult any textbook on graph theory, for example [3, 4, 7, 8, 9, 12, 18, 25, 27, 34, 37 and 41]. The following textbooks, [5, 10, 14, 24, 29 and 31], are devoted to problems concerning algorithms.

A *graph* is a pair $G = (V, E)$, where V is a set, the elements of which are called *vertices,* and E is a set of unordered pairs $\{u, v\}$ where u and v are distinct vertices in V. The elements of E are called *edges* of the graph. Throughout the book we assume that the set V is finite; infinite graphs are not considered. The sets V and E will also be denoted, for a given graph G, by $V(G)$ and $E(G)$, respectively.

Applications of graphs are almost inexhaustible. When we try to solve transport problems, we can consider edges as railways, or roads, and vertices as towns. In communication engineering, vertices may be transmitting stations or telephone exchanges. For the composition of school schedules, vertices may have the form 'history, 4th class' with edges connecting two vertices if they cannot be assigned the same time (for example, because the teacher is the same). In sociology, two persons of

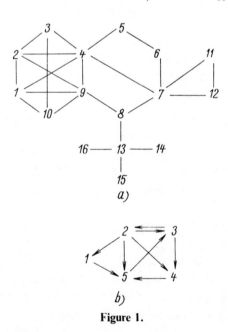

a)

b)

Figure 1.

a group under study will be connected by an edge if they like each other. In electrical engineering, the vertices may be radio components and edges the conductors, and so on.

Figure 1 is an example of a graph, the vertices of which are denoted by numbers, and the edges by line segments.

Given an edge $h = \{v, w\}$, we say that v and w are *adjacent* vertices, or vertices connected by an edge. The edge h is *incident* with the vertices v and w, which are the *endpoints* of h. For each vertex v of a graph G, the number of edges incident with v is called the *degree* of v, and this number is denoted by $\deg_G(v)$, or, more briefly, $\deg(v)$. More generally, let X be a set containing some vertices of the graph G. Then $\deg_G(v, X)$ or $\deg(v, X)$ denotes the number of edges with one endpoint v and the other lying in the set X. The sum of the degrees of all vertices of a graph G is equal to double the number of all vertices of G because each edge is counted twice (once for each endpoint).

A graph with vertices $v_1, v_2, ..., v_n$ can be represented by the *adjacency matrix* which is the square matrix $(a_{ij})_{i,j=1,...,n}$ of order n with the following entries:

$$a_{ij} = \begin{cases} 1 & \text{if } v_i \text{ is connected by an edge with } v_j, \\ 0 & \text{otherwise.} \end{cases}$$

The adjacency matrix is symmetric, its entries are 0 or 1, and it has zero on the diagonal. Conversely, each such matrix defines a graph. Other methods of presentation of graphs will be discussed in Chapter 3.

The complement of a graph $G = (V, E)$ is the graph (V, E') where E' is the set of precisely those pairs $\{v, w\}$ of vertices in V which fulfil $v \neq w$ and $\{v, w\} \notin E$.

Let us mention some important examples of graphs. For each set X we define the *complete graph* K_X on the set X to have as vertices precisely the elements of X, with each pair of distinct vertices connected by an edge. For each pair X and Y of disjoint sets we define the *complete bipartite graph* $K_{X,Y}$ on the sets X and Y as the graph with the set of vertices equal to $X \cup Y$ and with two vertices connected by an edge if, and only if, one lies in X and the other in Y. By a *circuit* is meant a graph with vertices $v_1, v_2, ..., v_n$ and the following edges:

$$\{v_1, v_2\}, \{v_2, v_3\}, ..., \{v_{n-1}, v_n\}, \{v_n, v_1\}.$$

Some authors admit loops (i.e. edges with coincident endpoints) or multiple edges, see for example [28], but we shall not use such generalisations.

By a *subgraph* of a graph G is meant graph G' such that $V(G') \subset V(G)$ and $E(G') \subset E(G)$. If all edges of G with both endpoints in the set $V(G')$

are edges of G', then G' is called a *full subgraph* of G. A full subgraph is uniquely determined by its set of vertices. For example, the full subgraph of the graph in figure 1a with vertices 7, 11 and 12 has edges $\{7, 11\}$, $\{7, 12\}$ and $\{11, 12\}$. If we delete the edge $\{7, 12\}$, we obtain a subgraph which is not full.

A *directed graph* G is a pair (V, E) where V is a finite set, called the set of vertices and denoted by $V(G)$, and E is a set of ordered pairs of distinct elements of the set V. The elements of the set E are called *directed edges*; if there is no danger of confusion, we say just edges. An example of a directed graph is illustrated in figure 1b, where the direction of each edge is denoted by an arrow. To distinguish the above two concepts, graphs are sometimes called *symmetric graphs,* or *undirected graphs*. Undirected graphs are often understood as special cases of directed graphs: each edge $\{v, w\}$ is represented by the pair (v, w) and (w, v) of directed edges. Thus results concerning directed graphs are, in this sense, more general than those concerning solely undirected graphs, since the former subsume the undirected case too.

A directed edge (v, w) has the opposite direction to the directed edge (w, v). A directed edge (v, w) leads from the vertex v and ends in the vertex w; v is its *initial* vertex and w its *terminal* vertex.

For a directed graph we can define the adjacency matrix by

$$a_{ij} = \begin{cases} 1 & \text{if there is an edge leading from } v \text{ to } w, \\ 0 & \text{otherwise}. \end{cases}$$

The adjacency matrix is not symmetric for a general directed graph.

By an orientation of an undirected graph $G = (V, E)$ is understood an arbitrary directed graph G' with the same set V of vertices, such that the following holds:

$$\text{if } (v, w) \in E(G'), \quad \text{then } (w, v) \notin E(G') \quad \text{and } \{v, w\} \in E(G)$$

and

$$\text{if } \{v, w\} \in E(G), \quad \text{then either } (v, w) \in E(G'),$$
$$\text{or } (w, v) \in E(G').$$

Orientation is not unique, and it must not be confused with G considered as a directed graph (which does not fulfil the first of the conditions above).

If a number is assigned to each vertex, or each edge of a graph, then we speak about a *vertex-labelled* graph, or *edge-labelled* graph, respectively.

By a *sequence* in an undirected graph G is understood a series of vertices $v_0, v_1, ..., v_k$ such that $\{v_{i-1}, v_i\}$ is an edge for each $i = 1, ..., k$. If the vertices are pairwise distinct, the sequence is called a *path*. The number k is called the *length* of the sequence or the path. If a path goes through all vertices of the graph, then it is called a *hamiltonian path*. If $v_0 = v_k$, we call the sequence *closed,* and if, in addition, $v_i \neq v_j$ for all $1 = i < j = k$, we call it a *circuit*. A *hamiltonian circuit* is a circuit passing through all vertices of the graph. Some examples related to the graph of figure 1a are:

a sequence: 1, 2, 3, 10, 3, 4, 9, 1, 2, 4, 7
a path: 1, 2, 9, 4, 5, 6, 7, 12, 11
a closed sequence: 7, 11, 12, 7, 8, 9, 4, 7, 6, 5, 4, 7
a circuit: 1, 2, 3, 4, 5, 6, 7, 8, 9, 10, 1.

Neither a hamiltonian circuit, nor a hamiltonian path exist in the graph of figure 1a.

A series $v_0, v_1, ..., v_k$ of vertices of a directed graph is called a *directed sequence* if for each $i = 1, 2, ..., k$, the pair (v_{i-1}, v_i) forms a directed edge. If the sequence is one-to-one, we get a *directed path*. The definition of a closed directed sequence is analogous to the undirected case; instead of directed circuit the name *cycle* is often used. A directed graph not containing any cycle is said to be *acyclic*. For the graph of figure 1b we have the following examples:

a directed sequence: 1, 5, 3, 2, 1, 5, 3, 4, 5
a directed path: 1, 5, 3, 2, 4
a closed directed sequence: 5, 3, 2, 5, 3, 4, 5
a cycle: 5, 3, 2, 4, 5.

A graph G is said to be *connected* if for two arbitrary vertices v and w there exists a path from v to w in the graph. A full, connected subgraph of a graph G which is not contained in any larger full, connected subgraph is called a *component* of G, or a connected component of G. Each graph can be uniquely decomposed into its connected components. A connected graph has a unique component, namely itself. The components of a graph are pairwise disjoint, and each vertex lies in one of the components. A connected graph is also said to be 1-connected.

A vertex of a graph is called an *articulation point* if the number of components of the graph is increased by deleting the vertex. For the graph of figure 1a, the articulation points are precisely the vertices 7, 8 and 13. A connected graph with no articulation points is said to be *biconnected*. A maximal full biconnected subgraph of a graph G is called a *biconnected component* of G. The biconnected components of the graph of figure 1a are determined by the following sets of vertices: $\{1, 2, ..., 10\}$, $\{7, 11, 12\}$, $\{8, 13\}$, $\{13, 14\}$, $\{13, 15\}$ and $\{13, 16\}$. Two biconnected components are either disjoint, or they meet in a single vertex. Each vertex lying in the intersection of two distinct biconnected components is an articulation point. Conversely, each articulation point lies at such an intersection.

More generally, an *n*-articulation of a graph G is a set X of n vertices such that by deleting those n vertices the number of components of G is increased. A 2-articulation is called a *biarticulation*. In figure 1a we have $\{4, 9\}$ and $\{8, 7\}$ as examples of biarticulation. A graph of more than n vertices is said to be *n*-connected if it is connected and does not have any *n*-articulation.

A full subgraph is called a 3-connected component if it either has just one vertex, or just two vertices connected by an edge, or it is 3-connected and it is not a part of any larger 3-connected full subgraph. The decomposition of a graph into 3-connected components is unique; if two 3-connected components meet, then their intersection consists either of an articulation, or a biarticulation (in both cases it has at most two vertices). Each graph is covered by its 3-connected components. In the graph of figure 1a the following full subgraphs are the 3-connected components: $\{1, 2, 3, 4, 9, 10\}$, $\{4, 5, ..., 9\}$, $\{7, 11, 12\}$, $\{8, 13\}$, $\{13, 14\}$, $\{13, 15\}$ and $\{13, 16\}$.

A directed graph is said to be strongly connected if for each pair of vertices u and v there exists a directed path from u to v, and a directed path from v to u. The graph in figure 1b is strongly connected. A maximal, full, strongly connected subgraph of a directed graph is called a *strong component*. The strong components are pairwise disjoint and they cover all of the graph.

By a *colouring* of a graph is understood a rule assigning numbers (or colours) to the vertices of the graph in such a way that the endpoints of each edge have distinct colours. The following table presents an example of a colouring of the graph of figure 1a by four colours:

vertex: 1 2 3 4 5 6 7 8 9 10 11 12 13 14 15 16

colour: 1 2 1 3 1 2 1 2 4 2 2 3 1 2 3 2.

The chromatic number of a graph G, denoted by $\chi(G)$, is the minimum number of colours needed for colouring G. The chromatic number of a graph in figure 1a is 4. A graph is said to be *bipartite* if its chromatic number is 2 or less. A graph is bipartite if, and only if, the set of vertices can be decomposed into two disjoint sets A and B such that each edge has one endpoint in A and the other in B.

An *independent* set of vertices is a set A such that no edge has both endpoints in A. The largest number of vertices of an independent set is denoted by $\alpha(G)$. Given a colouring of the graph, then each set of vertices having the same colour is independent. Consequently, in a graph G of n vertices, we have

$$\chi(G)\,\alpha(G) \geq n.$$

A *clique* is a set of vertices such that two arbitrary distinct vertices in it are connected by an edge. A clique of a graph is an independent set in the complement of G, and vice versa. The number of vertices of the largest clique of a graph G is denoted by $\gamma(G)$. It is obvious that for each graph G we have

$$\gamma(G) \leq \chi(G).$$

A *matching* of a graph G is a subgraph G' of G in which each edge has degree exactly 1. A matching is, therefore, represented by a collection of pairwise disjoint edges. The following is an example of a matching of the graph of figure 1a: $\{1,2\}$, $\{3,10\}$, $\{4,9\}$, $\{5,6\}$, $\{7,8\}$, $\{11,12\}$ and $\{13,15\}$.

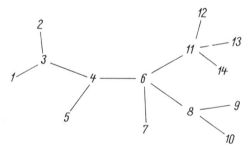

Figure 2.

A *tree* is a connected graph not containing a circuit. An example of a tree is depicted in figure 2. It is easy to prove that a tree of n vertices has exactly $n - 1$ edges, and each non-singleton tree has at least two vertices of degree 1. For each pair of vertices of a tree there exists exactly one path connecting the vertices. Given a tree of at least three vertices, then by deleting all vertices of degree 1 we obtain a non-empty graph which is also a tree. Repeating this operation, we reach the point at which only one or two vertices remain. If one vertex remains, then it is called the *centre* of the tree; if two, then they are called the *bicentre*. The two vertices of a bicentre are always connected by an edge. For example, in figure 2 we first delete the following vertices: 1, 2, 5, 7, 9, 10, 12, 13 and 14, and in the second stage, the vertices 3, 8 and 11, obtaining the bicentre: 4 and 6.

Let T be a tree, and let v be an arbitrary vertex of T. Then there exists a unique orientation of T such that for each vertex w we have a unique directed path from v to w. An example of such an orientation for the graph of figure 2, which has the described property for the vertex $v = 6$, is shown in figure 3a. The directed graphs obtained in the above manner

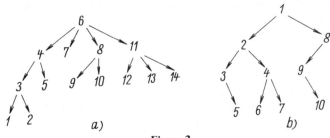

a)   b)

Figure 3.

are called *rooted trees*. The vertex v which was chosen as the basis of the orientation is called the *root* of the rooted tree. For each vertex w distinct from the root there exists a unique vertex z such that the pair (z, w) is a directed edge. The vertex z is called the *father* of w, and, conversely, w is called the *son* of z. (This terminology is derived in the obvious way from the masculine part of a family tree, and it can, equally obviously, be extended to additional concepts, e.g. uncle, brother, etc.) The root has no father. A vertex which has no son is called a *leaf*. Each rooted tree has at least one leaf. In figure 3a the vertex 8 is the father of the vertex 9, and the vertex 10 is a brother of 9 and the vertex 11 an uncle of 9, and so on.

Given a directed path $v = v_0, v_1, ..., v_k = w$ in a rooted tree, we say that v is an *ancestor* of the vertex w at height k above w. Conversely, w is in depth k below v. By the *depth* of a vertex is meant its depth below the root. By the *height* is meant the maximum of heights of the vertex over all leaves lying under it. The height of a rooted tree is the height of its root, i.e. the length of the longest directed path in the tree. All vertices of a given depth form a level of the rooted tree. In figure 3a the depth of the vertex 4 is 1, and its height is 2. The height of the tree is 3. The following is the list of all levels: 6; 4, 7, 8, 11; 3, 5, 9, 10, 12, 13, 14; 1, 2.

Figure 3a gives more information than that expressed by the concept of a rooted tree: the sons of each vertex are ordered from left to right. We shall often make use of such a structure which is called an *ordered rooted tree.* Thus, an ordered rooted tree is a rooted tree in which for each vertex there is given an ordering of the set of all sons. (To continue the analogy of the family tree, we say that a son is younger than all sons which succeed it in the given ordering.)

Figure 3b presents a rooted tree in a way that conveys still more information: for each vertex we can say whether it lies to the right or to the left from its father. This is typical for binary ordered trees, i.e. ordered trees in which each vertex has at most two sons (called the right and left son — naturally, either of these can be missing).

An important concept is that of the *drawing* of a graph **G**. This is a rule assigning to each vertex a point in the plane, and to each edge $\{v, w\}$ a connected, simple curve connecting the points assigned to the vertices v and w. If the drawing has the property that two arbitrary distinct curves in it have at most one common point, which then represents a vertex lying on both of the corresponding edges, then we say that the drawing is *planar*. In figure 4a we see a planar drawing of a graph. A graph

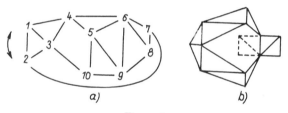

Figure 4.

which can be drawn in the plane in this way is said to be planar. The drawing of the graph of figure 1 is not planar; that graph does not admit a planar drawing, in fact. It can be proved that every tree has a planar drawing.

A planar graph together with a fixed planar drawing is called a *topological planar graph*. If all points lying on the curves of a topological planar graph, or representing its vertices, are deleted from the plane, the plane is decomposed into connected parts called the *faces* of the graph. One of the faces is unbounded, the others are bounded. A graph is planar if, and only if, it can be drawn onto a sphere without crossing of the curves representing edges. Here we can also consider the faces. The advantage of drawing onto a sphere is that the difference between bounded and unbounded faces vanishes.

The circumference of a face is bounded by the edges of a topological planar graph, or, more exactly, by the curves representing the edges. The number of edges constituting the circumference of a face is called its *degree*. We say that the face is incident with the edges lying on its circumference. For a topological planar graph we have a number of obvious geometric concepts. For example, given a vertex v we can move clockwise around v and thus obtain a cyclic ordering of all edges having the endpoint v. In this book we are going to use the concept of cyclic ordering of edges of a topological planar graph as a function q, assigning to each vertex v and each edge h with the endpoint v the first edge with the endpoint v which we meet when moving from the edge h clockwise around v. For example, in figure 4a for the vertex $v = 9$ and the edge $h = \{9, 10\}$ we have $q(v, h) = \{9, 5\}$, $q(v, q(v, h)) = \{9, 6\}$, etc. The cyclic ordering of edges is a highly suitable structure for a computer description of a topological planar graph. We shall investigate this in detail in § 3.9.

Given a planar drawing of a graph, it is easy to form its mirror image. If q and q' are cyclic orderings of the same graph corresponding to a planar drawing and its mirror image respectively, then $q(v, h) = h_1$ implies $q(v, h_1) = h$. In this way we can easily obtain the 'mirror image of information' concerning a planar drawing stored in a computer.

Given a connected topological planar graph of V vertices, H edges and S faces, Euler's theorem states that

$$V - H + S = 2.$$

This formula can be proved by induction: for small graphs it is easy to verify, and each connected graph can be expanded in two ways. Either we connect two existing vertices by a new edge, thus increasing the number of faces because the new edge will divide one face into two; this does not change the value of $V - H + S$. Or we add a new vertex, and connect it with an existing vertex by a new edge, which does not change the number of faces, and hence does not change the value of $V - H + S$ either.

Euler's theorem has a number of interesting consequences. If $V \geq 3$, then each face is incident with at least three edges. Since, furthermore, each edge is incident with at least two faces, we have $2H \geq 3S$, and consequently,

$$2 = V - H + S \leq V - H + \frac{2H}{3} = V - \frac{H}{3}.$$

Thus, $H \leq 3V - 6$. Inspecting separately the cases $V = 1$ and $V = 2$, we conclude that a connected planar graph of V vertices has at most $3V - 3$ edges. The same is also true for a disconnected graph because we can apply the rule to each of its components.

A planar graph of V vertices must contain at least one vertex of degree at most 5 because, if not, then the sum of degrees is more than $6V - 6$ which is double the maximum number of edges. Analogously, it can be shown that if the degree of each vertex is at least 3, then there exists a face of degree at most 5.

A *regular polyhedron* is a topological planar graph each face of which has the same degree and each vertex of which also has the same degree. One type of regular polyhedron is the *circuit,* all vertices of which are of degree 2. All other regular polyhedra have vertices of degree larger than 2, and hence, the degrees of both faces and vertices cannot be larger than 5. We conclude that, consequently, there are exactly five such regular polyhedra: cube, regular tetrahedron, octahedron, dodecahedron and icosahedron; the numbers of vertices of these polyhedra are respectively 8, 4, 6, 20 and 12.

Planar graphs are characterised by the Kuratowski theorem; we are going to use only a corollary of the theorem stating that a planar graph cannot contain a clique of five vertices.

It is relatively easy to prove that each planar graph can be coloured with five colours. One of the most outstanding problems, not only of graph theory but of all mathematics, was the so-called four-colour problem: is it possible to colour each planar graph with only four colours? Only recently this question has been answered affirmatively by K Appel and W Haken [46]. It is of interest that the solution made substantial use of a computer, which needed 1200 hours to find a collection of 2000 basic configurations appearing in planar graphs, and to present a colouring for each of them.

It can be proved that, given two planar drawings of a graph yielding the same cyclic ordering of the edges, then there exists a continuous transformation of the plane which maps the first drawing onto the second one. In §6.8 we shall ask what is the number of different cyclic orderings of edges of a planar graph obtained by various drawings of the given graph. We have already mentioned the possibility of performing the mirror image of a drawing, and the role this plays for the cyclic ordering of edges. In the case of a disconnected graph, we can take the mirror image of some components only and obtain new cyclic orderings in this way. In a connected planar graph with an articulation we can revolve some of the 2-connected components around the axis passing through the articulation. In a 2-connected graph with a biarticulation we can revolve some of the 3-connected components around a biarticulation, as indicated in figure 4b. In both cases we obtain new cyclic orderings of the edges. It can be proved that a 3-connected graph has only two

cyclic orderings of edges which are mirror images of each other, see
[210, 267]. In fact, by drawing a 3-connected graph in the plane, it can
be proved that the circumferences of its faces are exactly those circuits
which, when deleted from the graph, do not lead to a decomposition of the
graph into more components (verify this in figure 4a). A circuit also has
only two planar drawings, for obvious reasons, and these cannot be mapped
onto each other by a continuous transformation of the plane.

By an *isomorphism* of a graph $G = (V, E)$ onto a graph $G' = (V', E')$
is meant a bijection $f: V \to V'$ such that for arbitrary vertices x and y
in V we have that $\{x, y\}$ is an edge of the graph G if, and only if, $\{f(x), f(y)\}$
is an edge of the graph G'.

The definition of isomorphism for directed graphs is analogous. Two
isomorphic graphs are often considered as two copies of the same mathematical object, or they are even identified.

2

Models of computation

The reader has certainly encountered such concepts as computation, algorithm, and time or space needed by an algorithm. Exact and unambiguous definitions of these concepts are necessary for developing a mathematical theory of computation. This is the aim of the present chapter.

2.1 Random access machines

There exist a number of definitions of the concept of algorithm. Let us mention, for example, Turing machines [19], Markov normal algorithms [26], or recursive functions [35]. Furthermore, each program written in one of the high-level programming languages, such as ALGOL, FORTRAN, PL/I, or Pascal, is an algorithm. It has been proved that all these definitions are, in spite of their formal differences, equivalent. That is, any problem solvable by an algorithm according to one of these definitions (say, by a Turing machine) is also solvable by means of any other of these definitions (e.g. by a Markov algorithm). Moreover, Church's thesis states that the philosophical concept of 'algorithmic solvability' of a problem means precisely the solvability by means of a Turing machine (or a Markov algorithm, or an ALGOL program, or in any other of the equivalent ways).

The differences between the individual definitions of algorithm come to light, however, as soon as we ask not only which class of problems is algorithmically solvable, but also how much time an algorithm needs to find the solution. The algorithms used in theoretical considerations (Turing machines, etc) 'compute' in a way so different from that of real computers that results concerning the speed of computation are of rather limited practical value. On the other hand, an exact description of high-level programming languages is too complicated for the purpose of this book.

Models of computation which are relatively simple, and yet produce results which match real computations quite well, are based on computer programs at the level of assemblers, or machine codes. Although the differences between various individual machines at this level are considerable, the basic principles of programming are similar. In this book, we use one of the variants of the so-called random access machine, see [1] and [33]. A similar machine is also used in [22].

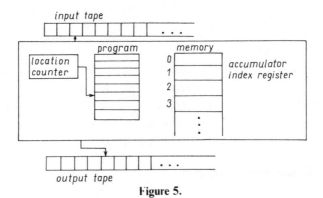

Figure 5.

The structure of a random access machine is shown in figure 5. The machine consists of a program unit, an arithmetic unit, a memory, an input unit and an output unit. The program unit contains a program consisting of a sequence of instructions which will be described below. It further contains a location counter showing which instruction is to be executed at a given instant. The arithmetic unit makes it possible to perform the basic arithmetic operations. The memory is divided into memory registers, each of which can hold an arbitrary integer. The memory registers are labelled by the numbers 0, 1, 2, 3, ... We assume, therefore, that there is no upper bound on the number of registers, or on their capacity; on the other hand, we shall always investigate both the number of registers needed by the algorithm under study, and the largest number stored in registers. The serial numbers allotted to the memory registers are called *addresses*. The registers with the addresses 0 and 1 have a special role, and they are called the *accumulator* and the *index register,* respectively.

The input data are conveyed to the machine via the input unit, consisting of an input tape and a read-only head. The input tape is divided into squares, each of which can hold an arbitrary integer. The head is able to read the contents of a square, and then it moves one square to the right. The structure of the output unit is similar, except that it has a write-only head which is able to print a number in a square of the output tape, and move one square to the right.

The name 'random access machine' refers to the ability of the machine to store information directly in memory registers, and to extract it from them, in the sense made clear below.

By a *configuration* of a random access machine is meant a rule assigning an integer to each square of the input and output tape, to each memory register, and to the location counter. (The numbers represent the contents of the squares, or the registers.) A configuration presents a complete description of the state of the machine at a given instant. The term initial configuration means a configuration such that the contents of all memory registers and all output squares are 0, the content of the location counter is 1, and for some non-negative integer n, the contents of all but the first n input squares are 0. This means that the computation starts with the execution of the first instruction of the program, with input data stored on the first n squares, while the rest of the machine is set to zero. An initial configuration is said to be determined by a sequence $c_0, ..., c_{n-1}$ of integers if the ith square of the input tape holds c_i for all $i < n$, and 0 for all $i \geq n$.

A computation of a random access machine is an infinite sequence $C_0, C_1, ...$ of its configurations such that C_0 is an initial configuration, and the passage from C_{i-1} to C_i, called a *step* of the computation, is determined by the program of the machine in a manner described below.

A *program* of a random access machine is a finite sequence $p_1, ..., p_n$ of instructions. The execution of a program is determined by the location counter, which holds the consecutive number of the instruction to be executed. After the termination of each step of the computation in which the instruction p_i has been executed, the content of the location counter is increased by 1. This means that the next instruction to be executed is p_{i+1}. There are several exceptions: the instructions *JUMP*, *JZERO* and *JGE* can change the contents of the location counter, and hence can direct the control of the computation to a location of the program different from the next instruction, and the halting instructions terminate the whole computation. It is possible, of course, for the program to go on forever, never executing a halting instruction.

We list below the instructions of a random access machine, divided into five groups.

Memory shift instructions

LOAD operand
STORE operand

Arithmetic instructions

ADD operand

SUBTRACT operand

MULTIPLY operand

DIVIDE operand

Input/Output instructions

READ

WRITE

Jump instructions

JUMP

JZERO label (jump if zero)

JGE label   (jump if greater or equal)

Halting instructions

STOP

ACCEPT

REJECT.

An operand determines either a number or an address on which the instruction operates. It can be presented in one of the following ways:

j, where j is a non-negative integer: j is the address of the operand, and the value of the operand is to be found in the memory register of the address j,

$*j$, where j is an integer: the address of the operand is $i + j$, where i are the instantaneous contents of the index register of the machine, and the value of the operand is to be found in the memory register of the address $i + j$,

$= j$, where j is an integer: the value of the operand is j, and no address is used.

If an instruction is executed with the operand $*j$, and if the address $i + j$ happens to be negative, it follows that there is an error of addressing, and the machine halts immediately.

A label is a natural number which determines the consecutive number of the instruction to be executed next in case a jump is made.

The individual instructions are executed as follows.

LOAD: the value of the operand is loaded in the accumulator, and the initial value of the accumulator is lost, while the rest of the memory remains unchanged.

STORE: the contents of the accumulator are loaded in the memory register, the address of which is equal to the address of the operand. The initial value of that memory register is lost while the rest of the memory (including the accumulator) remains unchanged. The instruction *STORE* cannot use an operand of the type $= j$.

Arithmetic instructions: the contents of the accumulator are changed, according to the type of instruction, either to the sum $a + b$, or the difference $a - b$, or the product $a \cdot b$, or the least integer equal to or greater than a/b, where a is the initial value of the accumulator and b is the value of the operand. An attempt to execute the instruction *DIVIDE* with an operand of value zero causes the machine to halt.

READ: the number currently read by the input head is stored in the accumulator, and the head moves one square to the right.

WRITE: the output head prints the contents of the accumulator in the current square of the output tape, and moves one square to the right.

JUMP: the value of the label is stored in the program counter. After this instruction, the location counter is not increased by 1. The label must not exceed the number of program instructions.

JZERO, JGE: if the contents of the accumulator are zero (for the instruction *JZERO*) or are greater than or equal to zero (for the instruction *JGE*), the value of the label is stored in the location counter. (This value must not exceed the number of program instructions.) If the accumulator does not satisfy the respective condition, no action is taken, and the contents of the location counter are increased by 1.

STOP, ACCEPT, REJECT: these instructions halt the machine. They have the same function in the course of computation, but we shall sometimes determine the result of the computation according to which of these instructions has been used.

For formal reasons, it will be convenient to look on the execution of the halting instructions as a step of the computation which does not alter the present configuration of the machine. Since the location counter is also left unchanged, the next step of the computation will execute the same halting instruction, and therefore, the configuration will remain unchanged. (We can easily imagine that the machine remains 'switched on' but nothing happens. This must not be confused with the well-known operation NOP = no operation, which, in spite of its suggestive name, increases the value of the location counter.)

The choice of instructions of a random access machine is substantially more limited than that of ordinary computers, but practically all of the currently used instructions can be obtained as a combination of a few of the instructions listed above.

2.2 Time and space complexity

When analysing algorithms presented in this book, we are interested to know not only whether we obtain a solution of the problem under consideration, but also how much time the computation is going to take and how much memory space must be reserved for it.

There exist a number of methods for estimating the speed of computation. The easiest one is to write the algorithm down in some programming language, to run it on a computer, and to determine the time taken by the computation. Naturally, an experiment like this can only be carried out for a limited number of test examples, and we cannot draw general conclusions from such tests. Besides, the results obtained in this way also depend on the technical parameters and program equipment of the computer, as well as on the choice of the programming language, and the capability and experience of the programmer. This is the reason why, in the present book, we concentrate on the theoretical analysis of algorithms understood as programs for random access machines. For some algorithms, however, results of computer experiments will also be presented.

One of the methods used to estimate the time needed for the execution of an instruction of a random access machine is based on the fact that for a binary expansion of a number n we need $\lceil \log(|n| + 1) \rceil \doteq \log |n|$ bits (plus the sign). The majority of instructions (except multiplication and division) can be executed in a running time proportional to the sum of lengths of the binary expansions of the numbers used in the instructions. That is, the sum of the lengths of their logarithms. The total running time of the computation is then determined as the sum of the durations of all of the executed instructions.

This logarithmic evaluation does not match the actual situation because the numbers we work with on a modern computer are usually stored in one or two memory registers, and the time needed for the execution of an operation on a memory register is nearly independent of the contents of the register. For this reason, a better agreement with empirical results is obtained by using the so-called uniform evaluation which presumes that the execution of each operation of the random access machine takes one time unit, and hence, the length of the computation is proportional to the number of the executed instructions. This is actually a simplification since, for example, $DIVIDE *j$ takes in practice more time than $JUMP$ or $STOP$ but the running time of the fastest and slowest instructions is constant, and hence the uniform evaluation makes it possible to determine upper and lower bounds for the actual length of the computation, accurate up to a multiplicative constant, which will be sufficient for our purposes.

Here, we shall use the uniform evaluation of the time and space complexity. This leads to the following definitions.

Definition 2.2.1. We say that a computation on a random access machine takes t time units if the tth step of the computation is either a halting instruction, or an address error, or a division by zero, whereas neither a halting instruction nor an error occurred on the ith step, for any $i = 0, 1, ..., t - 1.$ ●

Definition 2.2.2. We say that a computation on a random access machine uses memory of size m if some of the executed instruction used an operand of instantaneous address equal to m, whereas instantaneous addresses greater than m did not occur. ●

If the machine never halts, the time needed for computation is, according to the above definitions, undefined, and the memory size can also be undefined.

In the definition of a random access machine we have not given any limits to the size of a number to be stored in a memory register. It is a drawback of uniform evaluation that this inconsistency can be used to design seemingly fast algorithms, as the following example illustrates. The numbers 314, 159, 265, 358 and 979 can be concatenated into one number 314159265358979, and then both memory shifts and, to a certain degree, arithmetic operations on these numbers can be performed by one instruction, instead of five. When computing with matrices, whole columns or rows can be concatenated to achieve still more substantial 'reductions' of the time needed. We use quotation marks here because the execution of the program on a real computer requires storing large numbers in more than one or two memory registers and, therefore, no actual speeding up of the computation is achieved.

This deficiency is not going to invalidate the results of the more practically oriented Chapters 3 to 6, 9 and 10, but in the theoretical Chapters 7 and 8, we shall have to proceed more carefully, adopting the following restriction.

Restriction 2.2.3. Let P be a fixed polynomial. We admit only those computations of a random access machine which satisfy the following: if the computation is determined by the input data $c_0, c_1, ..., c_{n-1}$ then the absolute value of the contents of any of the memory registers in any of the computation steps does not exceed the number $P(\max(n, c_0, c_1, \ldots, c_{n-1}))$. ●

In the case of an attempt to perform such an instruction, we are going to assume that an error has been made, and the computation is terminated.

Restriction 2.2.3 applies also to the location counter, and hence to the amount of memory which can be used. Nevertheless, all problems solved in this book are solvable in polynomially bounded memory space (with trivial exceptions in Chapter 8) and thus we just eliminate pathological computations which would upset the proofs given in Chapters 7 and 8.

Should restriction 2.2.3 turn out to be too severe for a given problem, it is possible to store large numbers in several memory registers, and to perform arithmetic operations and transfers in the memory by suitable sequences of instructions (see the exercises at the end of the chapter).

On some occasions we need information concerning relations between the length of a computation and the amount of memory it requires. It is obvious that a computation of T steps does not use more than T memory registers. However, the required memory space, as defined in 2.2.2 above, is not given by the number of registers that have been used, but by the largest address of operand which states the size of the connected region of memory which is required. This value can be substantially larger than T. Although the cases in which the required memory is larger than the length of computation are exceptional, we cannot exclude them. We shall show, nevertheless, that each algorithm can be modified in such a way that its speed is not much decreased (from the theoretical point of view, at least), and the required memory space becomes smaller than the required time.

Theorem 2.2.4. *Let f be a function, and let* M *be a random access machine which computes each input sequence of length n in time at most $f(n)$. Then there exists a machine* M' *which computes each input of length n in time $O(f(n)^2)$ and with memory space $O(f(n))$, and which yields the same output sequence and terminates with the same instruction (or error indication) as the machine* M.

Proof. Let M' be a machine with two memory registers for each of the memory registers of address at least 2 on which M operates. Let us use the two registers of M' to store the address and the contents of the corresponding register of M, respectively. Such pairs of registers will be stored on the addresses $2i + D + 1$ and $2i + D + 2$ for $i = 0, 1, 2, \ldots$, and they will be followed (for denoting the end of the active part of the memory) by a register containing the number -1. The registers with the addresses $2, \ldots, D$ are auxiliary.

The machine M' first puts the number -1 into the register of address $D + 1$, and then it simulates the action of the machine M in the way described below.

An instruction of the machine M which performs a transition in the memory or an arithmetic operation and which does not have an operand of the form $= j$ is substituted in the machine M' by a sequence of instructions which guarantees that the machine acts as follows.

Let A be the address of the operand on which the simulated instruction of the machine M is acting. If $A < 2$, then the machine M' performs the same instruction as M. If not, the machine M' stores A in one of the auxiliary memory registers, and then it successively searches the memory registers with addresses $D + 1, D + 3, D + 5, \ldots$ until it finds a register with contents either A or -1. Let the register have address i. If the contents of the register are -1, then we first store the number A in the register, and then we store -1 in the register of address $i + 2$. In both cases, the instruction of the machine M is performed except that the address of the operand is changed to $i + 1$.

The other instructions are performed by M' in the same way as by M, except that for the jump instructions it is necessary to change the label appropriately. A detailed description of the simulation which would also allow us to determine the constant D is not difficult, although some technicalities have to be taken care of (e.g. the necessity to store the contents of the accumulator in an auxiliary register before performing the simulation of an instruction, and to return the contents to the accumulator before the final instruction of the simulating sequence). We leave the details to the reader as an exercise.

If the machine M requires $T = O(f(n))$ steps for the computation, then it used, beside accumulator and location counter, at most T memory registers and, therefore, the machine M' creates at most T corresponding pairs of registers. This implies an estimate of the memory space needed by the machine M'. The time of simulation of an instruction of the machine M is influenced in the first place by the search of the register with contents A or -1 (which, owing to the design of the procedure, will always be found). Due to the number of pairs of registers used by the machine M', the simulation of each instruction requires at most CT steps for a suitable constant C. ●

2.3 Problems and languages

A reader with some idea about the operation of computers will know that each problem can be presented to the computer in the form of a sequence of numbers. For example, the encoding of letters, numerals and other characters is internationally standardised, the logical values true and false are expressed by the numbers 0 and 1 respectively, etc. The notation used for more complex formations will be investigated in Chapter 3. It can also be assumed that the result of computation is given by the computer in the form of a sequence of numbers.

By the *concept* of a problem we can formally understand the following data:

1. a set A (usually infinite) of sequences of numbers, called the set of admissible input data, or the set of instances of the problem;

2. a rule U assigning to each sequence P in the set A a certain sequence of numbers $U(P)$ representing the result which corresponds to the given input data P. (Let us remark that misunderstandings often appear due to misuse of the concept of problem in situations where we have only a single instance of the problem or individual input data.)

It is our aim to construct a random access machine M which, given an admissible sequence P on its input tape, will return the sequence $U(P)$ on its output tape.

The analysis of the speed and memory space required by algorithms is performed as follows: we first determine, for each input data sequence, a number called the *size of input data*. The choice of that number is dependent on the type of objects described by the input data. It can be, for example, the length of the input sequence, or, in the case of graphs (regardless of the representation we use), the number of vertices, or the sum of the number of vertices and edges. The size of an $n \times n$ matrix is usually understood to be n (although n^2 numbers have to be ·given), and so on.

Then we determine what time the computation will take, and how much memory space it is going to use. We say that a random access machine M applied to a problem U requires time or space at most $f(n)$ if f is a function such that each input data sequence of size n is computed by M in time $f(n)$ or in memory space $f(n)$, respectively.

The evaluation of an algorithm in the above manner is called the *analysis of the worst behaviour* or the *worst case behaviour* because the results are valid even for the least favourable input data. In Chapters 3 to 9 we shall evaluate algorithms exclusively on the ground of their worst case behaviour because the results obtained are universally valid.

To simplify the analysis, we are going to determine the time or memory space consumption up to a multiplicative constant: we say that a given problem is solved by the given algorithm in time or space $O(f(n))$ if there exists a constant c such that the algorithm works in time or memory, respectively, at most $c f(n)$.

In Chapter 10 we will mention the analysis of the average, or expected behaviour, which determines the consumption of time or memory space as a mean, or a weighted mean, of the values obtained for all admissible input data of the given size.

With the exception of some parts of Chapters 7 and 8, we shall present a verbal description both of the input/output data and of the algorithms used because a formal description in some of the programming languages tends to stress the details of implementation inadequately, and to conceal the main ideas of the solution. We assume, however, that an experienced programmer who wants to use some of the algorithms presented will find no difficulties in creating a corresponding program based on our verbal description in some computer language, or even in the language of random access machines. For this reason we have not presented the above concepts of problem, input data, etc, in the form of numbered definitions.

In some of the more theoretically oriented parts of Chapters 7 and 8 it will, however, be desirable to use more exactly specified concepts. To avoid overcomplicated definitions, we introduce the following simplifications.

We are going to use input data exclusively as sequences of 0s and 1s. This will increase the consumption of the cells on tapes, but it will not cause any basic problems because numbers can be written in binary notation. Binary sequences will be called *words*. Furthermore, we shall mostly deal with problems of which the goal is to find out whether the given input data have a certain property or not (e.g. whether a given graph is connected) and, therefore, the desired output is 'yes' or 'no'. This makes it possible to use programs not containing the instruction *WRITE*, to exclude the application of the instruction *STOP*, and to determine the result of computation according to whether the final instruction is *ACCEPT* or *REJECT*.

When the above restrictions are accepted, then a presentation of a problem means a decomposition of the set of all words into two parts. The first part consists of all words which encode those input data for which the returned answer is 'yes', which means the words for which the corresponding computation terminates by the performance of the instruction *ACCEPT*, whereas the latter part, which is the complement of the first one, consists of codes of input data leading to the instruction *REJECT*.

Disregarding the fact that the input words often represent very complex mathematical objects, by a problem we can thus understand a finite or infinite set of binary words (which will correspond to the input codes for which the desired answer is 'yes'). Our aim is, then, to find a program of a random access machine which would accept an input word if, and only if, it belongs to the given set. Based on a linguistic analogy, sets of words are called *languages*. The concept of language will be used in the theoretical considerations below as a formalisation of a problem fulfilling the above conditions.

Definition 2.3.1. A *word* is a finite, possibly empty, sequence of 0s and 1s. The length of the sequence is called the *length* of the word. A *language* is a finite or infinite set of words. The set of all words is denoted by $\{0, 1\}^*$. ●

Detailed information concerning languages in the above sense can be found, for example, in [19].

The word 'word' will be used below in several meanings, but in Chapters 7 and 8 we are going to use it exclusively in the sense of definition 2.3.1.

Definition 2.3.2. An acceptor is a random access machine M the program of which does not contain the instructions *WRITE* and *STOP* and such that for each input word w the computation performed by M with the input sequence w stops after a finite number of steps by performing either the instruction *ACCEPT* or *REJECT*, without having made an address error, or by an attempt to divide by zero.

If the computation performed by the machine M with the input word w stops with the instruction *ACCEPT*, we say that M accepts w, otherwise M rejects w. ●

Thus, an acceptor either accepts or rejects each word. The accepted words form a language in the sense of definition 2.3.1; we call it the language accepted by the acceptor under consideration.

Definition 2.3.3. A transducer is a random access machine M such that the computation determined by an arbitrary input configuration has the following properties:

whenever the instruction *WRITE* is performed, either 0 or 1 is printed onto the output tape,
and
the computation stops after a finite number of steps by performing the instruction *STOP*.

The numbers written on the output tape in the course of computation determined by the input word w form a sequence denoted by $M(w)$. ●

Both an acceptor and a transducer must be designed in such a way that no computation can lead to an address error or a division by zero. An acceptor generates the language of all words it accepts, whereas a transducer leads to a self-map of the set $\{0, 1\}^*$. Definitions 2.2.1 and 2.2.2 above make it possible to measure the time and space complexity of functions and subsets (languages) of the set $\{0, 1\}^*$ as the time or space required by the best machine generating the given function or language.

Definition 2.3.4. Let $\mathbf{J}$ be a language and f a self-map of the set of all natural numbers. We say that the time or space complexity of the language $\mathbf{J}$ is at most f if there exists an acceptor M which accepts the language $\mathbf{J}$ and such that each computation of M determined by an input word of length n takes time or memory space, respectively, at most $f(n)$.

Exercises

1. Using the instruction set of random access machines, write down the part of a program which performs the jump to the instruction of serial number L in the case when the contents of the accumulator are less than zero, or less than or equal to zero. (Two and three instructions, respectively, are sufficient.)
2. Write down a program for a random access machine which, given an input sequence of the following numbers: $n, a_1, a_2, ..., a_n, b_1, b_2, ...,$ b_n will determine the value of the expression $a_1 b_1 + a_2 b_2 + ... + a_n b_n$ and will print it onto the first square of the output tape.
3. A quadruple of integers x_1, y_1, x_2 and y_2 printed on the input tape determines real numbers

$$r_i = x_i \cdot 10^{y_i} \qquad \text{for} \quad i = 1, 2.$$

Write a program for a random access machine which will compute the product $r_1 r_2$ and will print it onto the output tape in the form of two integers x_3 and y_3 such that

$$r_1 r_2 = x_3 \cdot 10^{y_3}.$$

4. Natural numbers A and B between 0 and 1 000 000 are encoded on the input tape of a random access machine in the form of four numbers a_1, a_2, b_1 and b_2 between 0 and 1000 such that

$$A = 1000 a_1 + a_2 \quad \text{and} \quad B = 1000 b_1 + b_2.$$

Write a program which will print onto the output tape the product $C = AB$ (or the sum $C = A + B$, or the integer part C of the quotient A/B) in the form of four numbers c_1, c_2, c_3 and c_4 between 0 and 1000 such that

$$C = 10^9 c_1 + 10^6 c_2 + 10^3 c_3 + c_4$$

in such a way that no instruction uses an operand with absolute value larger than 1000.

5. A graph of n vertices with the adjacency matrix (a_{ij}) is represented on the input tape of a random access machine by a sequence of the following type: n, a_{11}, $a_{12}, ..., a_{1n}$, a_{21}, $a_{22}, ..., a_{2n} ..., a_{n1}$, $a_{n2}, ..., a_{nn}$. Write a program which will print onto the output tape the number of edges of the graph and the degrees of each of the n vertices.

6. Determine the time and space complexity of the algorithms of exercise 5 above in dependence on n and the size of the given numbers.

7. If the input data of exercise 2 are printed in the following order: $n, a_1, b_1, a_2, b_2, ..., a_n, b_n$, then four memory registers are sufficient for the computation. Write down such a program.

8. Assume that no vertex of the graph G with n vertices has degree larger than 4. Design a way of describing the graph G on an input tape of a random access machine and write a program for the machine which would determine the number of edges of the graph G in time linearly dependent on n.

9. Let J be the language consisting of the words of the form a_0, b_0, $a_1, b_1, ..., a_n, b_n$, where n is a natural number, $a_0 = 1, ..., a_n = 1$, and the numbers $b_0, ..., b_n$ form a binary palindrome, i.e., $b_i = 0$ or 1 and $b_i = b_{n-i}$ for $i = 0, ..., n$. Write down a program for an acceptor which accepts the language J in time $O(n)$. (*Note*: the start bits $a_0, ..., a_n$ are necessary to enable the determination of the end of the sequence $b_0, ..., b_n$.)

3

Data structures

A considerable part of the computation time is often devoted to the manipulation of input and output data. The effectiveness of an algorithm thus often depends to a large extent on the way the manipulated data are organised. The requirements on data structures are usually quite contradictory, and it is necessary to find a compromise based on the character of the problem under study and the demands of the chosen algorithm. Keeping a lot of partial results will usually speed up the computation but it will increase the required memory space. Using more complex structures will make the computation shorter, but the programming will become more difficult. And so on.

Much effort has been devoted to the problems of data organisation both in their theoretical and practical aspects, and as a consequence, a vast literature concerning these topics can be found. The size of the present book does not permit more than a mention of the most important data structures, the selection of which has been motivated by the needs of the following chapters.

Most of the topics covered in the current chapter belong to the basic material which each programmer is supposed to know, and which can be found in a number of excellent textbooks or survey articles, e.g. [1, 6, 22, 23, 38, 40, 139 and 141].

3.1 Array and record

If we want to keep a number or a numerical piece of information in the memory of a random access machine, we proceed by reserving a memory register in which we shall store the number. Instructions manipulating the numerical value then make use of the address of the register.

We often need to keep and process not just one number but a system of data. In that case, we have to decide the following.

1. How are we going to store the data?

2. How are we going to describe which component of the data is to be processed, and how do we use that description to find the corresponding memory register, or the registers containing the required information?

One of the simplest methods for structuring a system of data is a (one-dimensional) array: we choose two numbers D and H such that the number of components of the array is equal to $H - D + 1$, and then we assign to each component one of the numbers $D, D + 1, ..., H$. The determination of components is then provided by specifying an integer I in the interval $D \leq I \leq H$, called *index*. For example, if the name of the array is P, and if we choose $D = 1$ and $H = 100$, the components of the array P are denoted by $P(1), P(2), ..., P(100)$. If the components of the array are integers, the usual way of storing them in the memory is to choose a number A larger than $-D$ and to store the component of the array of index value I in the memory register of the address $A + I$. Then the array lies in the memory registers of the addresses $A + D, ..., A + H$.

The localisation of the component of the array determined by the index is easily performed by a random access machine: we can, for example, store the value of the index in the index register and then execute instructions with the address A modified by indexing (e.g. $LOAD * A$).

It is often admitted that the components of the array represent a data system with an individual structure. It is required, however, that all the components have a data structure of the same type and the same size. Storing data in the memory of the random access machine is then usually executed in such a way that the following requirements are met: each component occupies a connected region of the memory, and if storing one component requires K registers then the component determined by the index I is stored in the registers

$$A + K * I + J \qquad \text{for} \quad J = 0, ..., K - 1,$$

where A is an integer chosen in such a way that $A + K * D \geq 0$.

A two-dimensional array is a generalisation of the above method of structuring data. Components of the array (which, again, are all of the same type) are determined by pairs of integers I_1 and I_2, satisfying the following inequalities:

$$D_1 \leq I_1 \leq H_1 \quad \text{and} \quad D_2 \leq I_2 \leq H_2$$

where D_1, D_2, H_1 and H_2 are chosen constants. If each component is stored so as to occupy K successive memory registers, then the component determined by the indices I_1 and I_2 is stored in the memory region beginning with the address

$$A + K * (I_1 * (H_2 - D_2 + 1) + I_2)$$

where A is a constant chosen in such a way that the above expression cannot be negative. Similarly, we can express an array of any arbitrary dimension.

It is easily seen that for each type of array, given by the dimension, the bounds of the indexes and the type of the components, it is possible to localise each component in a constant amount of time, independent of the magnitude of the indices (assuming, of course, that the values of the indices are known — otherwise the whole operation is prolonged by the computation of those values).

Another simple way of data structuring is *record*. A record often consists of a single component, e.g. an integer. More complex records have components determined by names. For example, when describing vertices of a directed graph, we may use a record with components called *ORD* (the ordinal number of the vertex), *DEG* (the degree of the vertex), and *VAL* (another numerical value specified for the vertex). Calling the item *VER*, the components are usually denoted by *VER.ORD*, *VER.DEG* and *VER.VAL*, respectively.

Assuming that each item of the record is stored in a connected region of the memory, it is useful to place the record in the memory in such a way that the components follow one another directly. If we know the names of the components, their ordering, and the manner in which they are represented (in particular, the area of the memory region which they occupy), it is sufficient, for the localisation of each component, to know its name and the address of the record, i.e. the address of the first memory register used to store it; the time needed for this is again constant.

We shall often deal with data structures the elements of which carry an individual structure of records of the same type. Sometimes records forming a data structure have the following property: one of the items distinguishes the individual records in the sense that distinct records have distinct values of the item. Such an item is called a *key*. Each record contained in the structure is uniquely determined by the value of its key. For example, if vertices of a graph are described by the set which for each vertex contains one record with items *ORD*, *DEG* and *VAL* (see above), then the ordinal number *ORD* is certainly a key whereas the component *DEG* is not a key because we often find two vertices in a graph with the same degree.

In subsequent sections of this chapter we shall often study the problem of implementation of various data structures in such a way that the search for an element determined by the value of a key is as simple and, above all, as quick as possible.

3.2 Lists

It is very often necessary to store in the memory of the computer a sequence of certain objects, e.g. variables or records, for which the ordering in the sequence is important. Here we allow the same object to be repeated in the sequence. When speaking about data structures, sequences are often called *lists*.

The basic operations on lists are the following:

determine whether a given item is on the list,
add a given item at the beginning or the end of the list,
insert a given item between two chosen elements of the list,
delete an item from the beginning or the end of the list,
delete an arbitrary item which is an element of the list,
concatenate two lists,
split a list into two,
determine the number of elements on the list.

The speed with which it is possible to perform the individual operations depends on the implementation of the list, i.e. on the manner in which it is stored in the computer. The requirements of simplicity of implementation and speed of performance of the operations are usually contradictory and, moreover, speeding up one operation can lead to slowing down another. The speeding up of operations is also obtained at the cost of the required memory space. It is therefore necessary to consider which operations will actually be performed during the computation and avoid designing too general and, consequently, overcomplicated implementations.

The following special cases of lists are often used.

Stack which permits only addition and deletion of the last item on the list, and the determination of the last item. Stack is also called last-in-first-out. (It is sometimes assumed that deleting and inserting takes place at the beginning rather than the end of the list.)

Queue which permits only insertion at the end and deletion from the beginning, and determination of the item at the beginning of the queue. This is also called first-in-first-out.

Two-sided queue in which we can add, delete and determine items at both ends.

We shall now present two of the basic ways of implementation of lists.

A. Compact lists

This method can be used when the elements on the list are objects of the same type. We proceed by choosing a one-dimensional array with index values from 1 to L_{max} which has at least as many components as the maximum number of items contained in the list under consideration. The kth element of the list will be stored in the component of the array determined by the index k. Moreover, we shall use a variable determining the number of elements of the list.

If the name of the array is, for example, S and the length of the list is given by the variable L, then the elements of the list lie in the successive components $S(1)$, $S(2)$, ..., $S(L)$. We shall now show how the basic operations are performed on such a list.

Algorithm 3.2.1. Addition of a new item at the end of the list

1. [*Expanding the list*] $L := L + 1$.
2. [*Adding an item*] $S(L) :=$ the item to be added.   ●

It is useful to supplement step 1 of the algorithm by a check on whether the length of the list has not exceeded the permitted length given in the declaration of the array. This applies also to all the following algorithms in which the length of the compact list is increased.

Algorithm 3.2.2. Addition of a new item after the *I*th element on the list $(0 \leq I \leq L)$

1. [*Expanding the list*] $L := L + 1$.
2. [*Transferring items*]   For $J := L$, $L - 1$, ..., $I + 1$ do $S(J) := S(J - 1)$.
3. [*Adding an item*] $S(I) :=$ the item to be added.   ●

Algorithm 3.2.3. Deletion of an item from the end of the list

1. [*Shortening the list*] $L := L - 1$.   ●

Shortening is meaningful only if the list is non-empty. When this operation is performed, it is useful to check whether the list is non-empty.

Algorithm 3.2.4. Deletion of the *I*th item from the list

1. [*Shortening the list*] $L := L - 1$.
2. [*Transferring items*] For $J := I, I + 1, ..., L$ do $S(J) := S(J + 1)$. ●

Algorithm 3.2.5. Concatenation of a list *S2* of length *L2* to a list *S1* of length *L1*

1. [*Transferring elements*] For $J := 1, ..., L2$ do $L1 := L1 + 1$; $S1(L1) := S2(J)$. ●

Algorithm 3.2.6. Creation of a list *S2* by splitting from the list *S1* the part beginning at the *I*th element

1. [*Initiating S2*] $L2 := 0$.
2. [*Transferring elements*] For $J := I, ..., L1$ do $L2 := L2 + 1$; $S2(L2) := S1(J)$.
3. [*Shortening S1*] $L1 := I - 1$. ●

It is clear that adding and deleting at the end of the list and determining the number of elements are fast operations the duration of which is independent of the number of elements in the list. In contrast, adding and deleting a general element, as well as concatenation and splitting of lists, are time consuming operations because in the worst case they require the transfer of a large part of the elements of the list.

Consequently, using compact lists is suitable primarily in the case of stacks; otherwise they are used only when the length of list is very small.

The method of implementation of compact lists can, however, be modified in such a way that it becomes suitable also for a queue. If we do not transfer the elements forward after deleting the first element, the queue is going to 'retreat'. We shall therefore use two variables, denoted by B and E, which will describe the beginning and the end of the queue. If the name of the array is S, then the beginning of the queue is $S(B)$, and the end is $S(E)$. At the instant at which the value of E would exceed the upper bound L_{max} of the value of the index of the array S, we are going to add the elements of the queue at the beginning of the array, i.e. starting with the component $S(1)$. This situation is indicated by the validity of $E < B$, and the elements of the queue are then $S(B), ..., S(L_{max})$, $S(1), ..., S(E)$. During the further transfer of the queue it will happen after some time that the beginning will also jump to the beginning of the array S, and thus, the normal situation with $L \leq E$ will return, and so on.

The above method can also be applied to a two-sided queue the end of which, however, can move in both directions.

Queues with 'floating' ends make it possible to perform adding and deleting on both ends, in a time independent of their lengths.

Besides the time complexity, compact lists have another serious disadvantage which will be illustrated by the following example. We are to create 100 lists which together will constitute 1000 items. Although the average length of a list is 10, we have to be prepared for the possibility that all items will be concentrated in one list. We consequently have to use 100 arrays per 1000 components which is a great waste of memory, or to distribute memory dynamically in the course of computation which can lead to a great waste of computation time. This is the reason why a much more flexible method of implementation of lists is in use, which we are going to describe presently.

B. Linked lists

The individual items are going to be distributed in an arbitrary manner in the memory, but each item will previously be expanded by one integer variable which will specify the address of the next item on the list. The

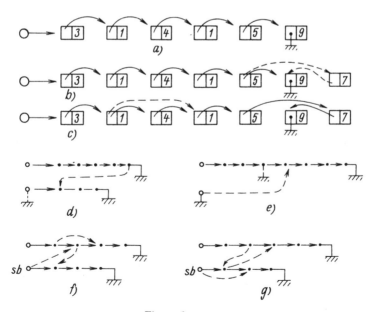

Figure 6.

number which is the value of that variable is called a *pointer*. The description of the structure is supplemented by a variable called the *heading* which will specify the address of the first item. If the distribution of items in the memory is performed by a high-level language compiler, then the actual value of the pointer is not important for a programmer, only the fact that it points to the address of the next item (which, moreover, can have d'fferent values during repetitions of the same computation). Some of the high-level languages, e.g. PL/I and Pascal, therefore introduce a special type of variable for manipulation with pointers.

Figure 6a presents a list containing the numbers 3, 1, 4, 1, 5, 9. An example of their distribution in the memory of a computer is seen on the first two rows of the table below: each item is stored in two successive memory registers the first of which contains the pointer and the latter the represented item. The address of the heading of the list is 10 and the pointer of the last item is equal to 0 (registers 22 and 23). The registers of addresses 11, 14, 15, 20 and 21 are unused.

Address:	10	11	12	13	14	15	16	17	18	19	20	21	22	23	24	25	26	27
Contents of register:	16	14	18	1	20	37	12	3	24	4	0	21	0	9	26	1	22	5
Adding 7:	16	20	18	1	22	7	12	3	24	4	0	21	0	9	26	1	14	5
Deleting 4:	16	18	24	1	22	7	12	3	20	4	0	21	0	9	26	1	14	5

Adjoining an item of value 7 after the item of value 5 is illustrated by figure 6b and the third row of the above table. The newly adjoined item is placed in an unused register and the corresponding re-connection of the pointers is performed. Concatenation and splitting of lists is illustrated in figures 6d and 6e. If concatenation is often performed, it is useful to introduce a variable the value of which is the position of the last item on the list.

Although the operations on items are very simple and rapid, certain difficulties can occur.

The first problem is searching a free memory register for storing new items, and disposing of the place left by deleted items. The simplest method is the carefree one: we occupy every time the free positions in the memory, and we leave unused the positions made free by deleting items. The programmer who wants to prevent his program from suffocating in its own garbage can use the so-called collector. This is an auxiliary linked list which collects all the unused memory registers. After deleting an item from the main list we immediately place it into the collector (preferably at the beginning), thus keeping it ready for further use, see figure 6f. If, on the other hand, we add a new item to the main list, we delete some item from the collector (preferably the first one, again), storing the new item there and then adding it to the main list, see figure 6g. In the above table we used a garbage collector with the heading of address 11 which initially contains two items with the register addresses 14, 15, and 20, 21. Verify the function of the collector!

At the beginning of a computation it is useful to partition the memory which is at the disposal of the list into sections of lengths corresponding to the lengths of the items, and to order those sections into the garbage collector.

Another complication connected with the application of linked lists occurs when items are deleted. For this operation it is necessary to determine the predecessor Z_1 and the successor Z_2 of the item to be deleted. The determination of the successor is easy because it is described by the pointer of the item considered. The determination of the predecessor is, however, complicated because in general we must search the whole list from the heading until Z_1. (The first item is an exception because its predecessor is the heading.) For this reason, doubly-linked lists are used, in which we add two pointers to each item, one with the address of the predecessor and the other one with that of the successor, see figure 7a. Deleting from the list is illustrated in figure 7b and adding in figure 7c.

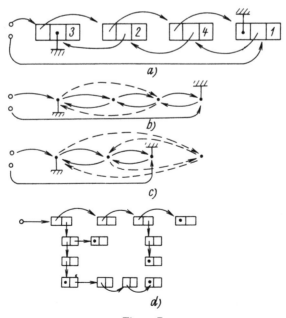

Figure 7.

The following table illustrates a way of describing the situation of figure 7a: the heading has the address 10, and another variable is used with the address 11 which points to the last item on the list. Items occupy three successive memory registers: the pointer to the first register of the successor, the pointer to the second register of the predecessor, and the actual contents of the item. The collector is not used. The first and second line describe the list 3, 2, 4, 1 from which the item 4 is afterwards deleted (see the third line of the table), then the item 7 is added after the item 3 (see the fourth line), and finally, the item 1 is deleted (see the fifth line).

Address:	10	11	12	13	14	15	16	17	18	19	20	21	22	23
Contents:	18	22	15	19	2	21	13	4	12	0	3	0	16	1
Deleting 4:	18	22	21	19	2	21	13	4	12	0	3	0	16	1
Adding 7:	18	22	21	16	2	12	19	7	15	0	3	0	16	1
Deleting 1:	18	13	0	16	2	12	19	7	15	0	3	0	16	1

Besides the methods mentioned for constructing linked lists, it is also possible to construct cyclic lists in which the pointer of the last item points to the first item on the list, or various branched structures in which the elements of the main list are headings of further linked lists, as illustrated in figure 7d. The last mentioned possibility will be used for an implemen-

The main advantage of linked lists is the possibility of making full use of the memory at our disposal, since one region of the memory can contain several lists at the same time and, conversely, the items of one list can be placed quite arbitrarily in various parts of the memory. Application of doubly-linked lists makes it possible to perform all the operations mentioned at the beginning of this section in a constant time, independent of the number of elements of the list, with the exception of determining whether a given item is on the list. This can also be achieved by a singly-linked list, provided that the required menu of operations to be performed is smaller (as in the case of stack or queue).

A disadvantage is that by using auxiliary variables (pointers) we decrease the amount of the memory available for the main data.

3.3 Rooted trees

A suitable implementation of rooted trees is important in its own right and, moreover, we shall see that it plays an important role in the effective implementation of sets. The methods we present describe, in fact, ordered rooted trees. For a representation of ordinary trees, it is sufficient to choose an arbitrary ordering of the sons of vertices, and then apply the same methods. To simplify terminology, we assume that the sons of each vertex are ordered according to their 'age'.

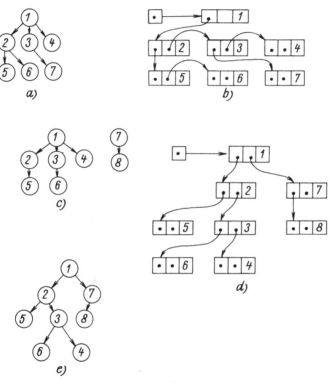

Figure 8.

One possibility of representation of an ordered rooted tree is to use for each vertex an item containing, *inter alia*, two pointers: one points to the address of the oldest son, and the other one to the address of the eldest of the younger brothers. If one of those vertices does not exist, the corresponding pointer contains a value distinct from any address. An example of a representation of the tree in figure 8a is shown in figure 8b. The description also contains a heading showing the root of the tree.

For a representation of a forest, the same method is applied with an additional ordering of the roots of the individual trees into a linked list. For this, we can make good use of the free pointers to the younger brother, since the roots do not have brothers. An example of a representation of the forest of figure 8c is shown in figure 8d.

For a representation of a binary tree it is suitable to use two pointers for the description of each vertex containing the addresses of the right and left son (respectively, a special value indicating their non-existence). An example is shown in figures 8e and 8d.

The above methods of implementation lead to an interesting correspondence between rooted forests and binary trees: a forest is represented using pointers to the eldest son and younger brother, or to the next root, and the resulting description can be considered as a binary tree with the method of pointing to the right and left son. We can also proceed in the reverse direction, from a binary tree to an ordered rooted forest. An example of this correspondence is illustrated in figures 8c, d, e.

Analogous to the backward pointer of a doubly-linked list we can add to the representation of rooted forests a pointer to the father. This is used mainly in situations in which we have to proceed in the direction towards the roots.

The basic operations, e.g. adding a successor of a given vertex, or deleting a vertex or a subtree determined by a vertex, are performed by a manipulation of the pointers in a manner similar to that for linked lists. The concrete performance of these operations is left to the reader as an exercise.

A suitable modification of the above procedures using garbage collectors of the deleted vertices can also be applied.

3.4 Sets

In the present section we study data structures designed for a description of a set, or a collection of sets. We first list some of the operations performed on sets.

$MEMBER(x,M)$: determine whether x is an element of the set M, and usually determine also the position in the memory where x is stored when $x \in M$.

$INSERT(x,M)$: add x to the set M. This operation is sometimes implemented in such a way that for its proper function it is necessary that x does not lie in M. If we are not sure about this, we must first perform the operation $MEMBER(x,M)$.

$DELETE(x,M)$: delete x from the set M. We usually assume that x actually lies in M and that its address is known. If this information is not at our disposal, we must first perform $MEMBER(x,M)$.

Further operations are based on the fact that elements of sets are items which can be compared by keys. For simplicity, instead of the following formulation 'the key of the item x is larger than the key of the item y' we shall simply say 'the item x is larger than the item y'.

The operations based on comparison are the following:
$MIN(M)$: determine the item in M with the least value of the key.
$EXTRACT\text{-}MIN(M)$: determine the item in M with the least key, and delete it.

Analogously, we can introduce the operations $MAX(M)$ and $EXTRACT\text{-}MAX(M)$. The operations $EXTRACT\text{-}MIN$ and $EXTRACT\text{-}MAX$ are actually combinations of the operation MIN or MAX, and $DELETE$. They are useful when the data structure does not make it possible to delete easily an arbitrary element of the set, only an element with an extreme key value, stored in a suitable position of the structure.

Comparing keys of items is also used for the implementation of the operations $MEMBER$, $INSERT$ and $DELETE$ because the comparison makes it sometimes possible to achieve a substantial increase of the speed of those operations.

We shall sometimes, e.g. in Chapter 5, use data structures for description of sets permitting a repetition of the same item at two different positions, or, at least, two items with the same key. The following modification of the operation $INSERT$ is then meaningful:

$= INSERT(x, M)$: let the key of the item x be k. If no element of M has key k, proceed as with the operation $INSERT$. Else, proceed by inserting x using $INSERT$ with a slightly increased key of x (just so as to remain smaller than the key of any member of the set M having the key larger than k).

When we are storing information concerning several sets, we can perform the usual operations such as union, intersection, etc. We often restrict ourselves to unions of disjoint sets because then in the resulting set we do not have to delete a repeated occurrence of some of the elements. In this book, we are going to use two operations on systems of mutually disjoint sets:

$FIND(x)$: determine in which set of the disjoint system the element x lies (the disjointness guarantees the uniqueness of the result).
$UNION(M, N)$: form the set-theoretical union of the disjoint sets M and N.

We saw in §3.2 that the types of data structures with a restricted menu of operations can be implemented in a simpler way. In this book, the following types of data structures will be of primary importance for implementation of sets:

Name	Operations
Dictionary	INSERT, DELETE, MEMBER
Priority queue	MIN, INSERT, DELETE
Heap	MIN, INSERT, EXTRACT-MIN
Mergeable heap	MIN, INSERT, DELETE, UNION
Factor set	UNION, FIND

We now turn to the simplest methods of implementation of sets.

A. Implementation by means of the characteristic function

This method is used when the set under consideration is a subset of some basic set M_0. The elements of the basic universe M_0 will be denoted by $x_1, ..., x_n$, and for a description of the set M we shall use an array $A(i)$, $i = 1, ..., n$, the components of which are logical variables (or numbers 0, 1): $A(i) =$ true is interpreted as $x_i \in M$, and $A(i) =$ false as $x_i \notin M$. The empty set is thus obtained by putting $A(i) =$ false for $i = 1, ..., n$. The operations MEMBER, INSERT and DELETE can be performed in a constant time independent of n and of the size of M, because $MEMBER(x_i)$ is just the determination of the value $A(i)$, and the other two operations are obtained by setting $A(i) =$ true and $A(i) =$ false, respectively.

The union and intersection of sets represented by arrays A and B is performed by the determination of the logical sum and product, respectively, of the values $A(i)$ and $B(i)$ for all i.

When performing the operation MIN, it is useful to order the elements of the universe M_0 in such a way that $x_1 < x_2 < ... < x_n$. Then it is sufficient to search the values of i from 1 until $A(i) =$ true. In the worst case, however, the time required by this procedure is proportionate to n. P van Emde Boas has found an ingenious method which makes it possible to perform the operations MIN, MAX, INSERT and DELETE and to determine the predecessor and the successor in time $O(\log \log n)$, and to perform the operation MEMBER in a constant time independent of n; see [261, 262].

Implementation using the characteristic function is particularly useful when the size of the basic universe M_0 does not exceed the number of bits in one word of the computer (which is usually 32 or 16) since then it is possible to code the complete information about M into a single word, and the union and intersection are then performed by a single instruction. On the other hand, this implementation is impractical if M_0 is too large (e.g. the set of all numbers which can be represented in the computer, or the set of all identifiers of some programming language), even if M itself is small.

Implementation using the characteristic function is usually found in the data type 'set' of the language Pascal.

B. Implementation of a set as a list

We can order the elements of the set in any way, and then use the methods described in §3.2.

Performance of the operation *MEMBER* requires searching all of the list, or the substantial part of it. The operation *MIN* requires searching the whole list in any case. Deleting an element of a known address can be performed in a constant time independent of the size of the set provided that the list is implemented as a doubly-linked list. The operation *INSERT* is even simpler, because the new element (provided we are sure that it does not belong to the set) can be added into an arbitrary position in the list, e.g. at the beginning, or at the end.

The union of disjoint sets is performed by concatenating the corresponding lists, and hence, when linked lists are used, this operation is simple and rapid. The general union, in contrast, has large time complexity because it requires searching and deleting elements which appear twice in the union. Intersection of general sets is also difficult to perform.

C. Implementation of a set as an ordered list

This method requires making use of keys of the items which form the elements of the set. The items are ordered so that the key values increase and then some of the methods of §3.2 are used. In this case, the performance of the operation *MIN* is particularly simple, and the same is true of *EXTRACT-MIN* provided that suitable implementation is used, because the element with the least key lies at the beginning of the list. When doubly-linked lists are used, the operation *DELETE* is also very rapid when the address of the element to be deleted is known. In contrast, the time complexity of the operation *INSERT* is large because it requires first finding the position where the added element will be stored, provided that the ordering of the list is to be preserved.

The operation *MEMBER* requires searching the whole list or a substantial part of it. An exception is the implementation of the set as a compact ordered list since then *MEMBER* can be speeded up by bisection. That is, if we know that the searched item is in the given array S between $S(B)$ and $S(E)$, then we denote by I the integer part of the quotient $(B + E)/2$, and by comparison of the item under search with $S(I)$ we determine whether the item is equal to $S(I)$, or lies in $S(J)$ for $J = B, ..., I - 1$, or in $S(J)$ for $J = I + 1, ..., E$. If this has not led to a localisation of the item, the region of the possible occurrence is diminished to a half. By repetition of this procedure it is possible to perform *MEMBER* in time $O(\log n)$ where n is the number of elements of the set. Unfortunately, the implementation as a compact list does not make it possible to perform quickly the operations *INSERT* and *DELETE*, and therefore this method is suitable only when the contents of the set are not often changed, and in a majority of instructions searches are only performed by means of *MEMBER*.

In the subsequent sections of the present chapter we are going to describe more rapid, although occasionally substantially more complicated, methods of implementation by means of dictionary, priority queue, heap and factor set.

3.5 Search trees

This and the subsequent section will deal with more complicated structures for implementation of sets based on an application of rooted ordered trees, which make it possible to perform the majority of operations in time proportionate to the logarithm of the number of elements of the set. For simplicity, we shall assume that the elements of the sets described are not general items but integers.

In the present section we mention search trees which are suitable above all for a representation of dictionaries and priority queues. The simplest among them is the binary search tree.

Definition 3.5.1. A *binary search tree* is a binary tree whose vertices are labelled by integers in such a way that the following holds for each vertex: if y is the left or right son of the vertex x, then neither y nor any direct or indirect successor of y has label larger, or smaller, respectively, than x. ●

For a description of a set of n elements we can use an arbitrary binary tree of n vertices which are labelled by the elements of the set under description in such a way that the condition of definition 3.5.1 is satisfied. Figure 9a illustrates how the following set, 1, 2, 3, 5, 7, 8, 10, 11, 14, 15, 18, is stored.

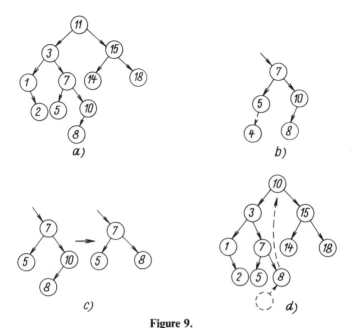

Figure 9.

The above method of labelling the vertices of the tree considerably simplifies the performance of operations. We first show the procedure for the operation *MEMBER*.

Algorithm 3.5.2. Search for a vertex with label *C* in a binary search tree with labelling given by a function *c*

Auxiliary variable: v, the vertex being investigated.

1. [*The beginning of the search*] If the tree is empty, the searched item naturally does not lie in it. Else, set v equal to root of the tree.

2. [*Procedure*] The following operations are cyclically repeated:

2a. if $C = c(v)$ then v is the vertex we are searching for, and the computation is terminated;

2b. if $C < c(v)$ and the vertex v does not have a left son, then the vertex we are searching for does not exist, and the computation is terminated;

2c. if $C < c(v)$ then set v equal to the left son of v;

2d. if $C > c(v)$ and the vertex v does not have a right son, then the vertex we are searching for does not exist, and the computation is terminated;

2e. if $C > c(v)$, then set v equal to the right son of v. ●

For example, searching for a vertex with label 10 in the tree in figure 9a, we pass through the vertices 11, 3, 7, 10, whereas the unsuccessful search for a vertex with label 6 would pass through the vertices 11, 3, 7, 5, and a further step to the right son of 5 is not possible.

The operation *INSERT* is equally simple.

Algorithm 3.5.3. Addition of a vertex with label C to a binary search tree
We proceed as in algorithm 3.5.2 with the following modifications.

If the given tree is empty, we create a tree of a single vertex with label C. If a vertex with the label C is found in the original tree, then the computation terminates without adding another vertex.

If the computation according to algorithm 3.5.2 would be terminated in steps 2b or 2d, then a left son or right son, respectively, is added to the last vertex v, with label C. ●

Figure 9b illustrates the addition of a vertex labelled 4 to the tree in figure 9a.

The binary search tree makes it possible to perform the operations *MAX* and *MIN* very easily.

Algorithm 3.5.4. Operations *MIN* and *MAX* in the binary search tree
Set $v :=$ the root of the tree, and then perform the operation $v :=$ the left (or right, respectively) son of v as long as the son under consideration exists. When a vertex is reached in this way, which does not have the left son (or right son, respectively), then we have found the vertex with the extremal value of the label, and the computation is terminated. ●

The operation *DELETE* can turn out to be somewhat more complicated.

Algorithm 3.5.5. Deletion of a vertex v from a binary search tree
1. [*v has no sons*] If v has not got sons, we delete it from the tree.
2. [*v has one son*] If v has one son w then, in the case when v is the root, we delete v from the tree and declare w as the new root; if v has a father u, we delete v from the tree and we connect w as a new left son or right son of u, according to whether v was a left son or a right son of u.
3. [*v has two sons*] Let v have both sons, and denote by w the left son. Put $z := w$ and then repeat the instruction $z :=$ the right son of z as long as z has a right son. Then the label of z is transferred to y, and the vertex z is deleted by the methods of steps 1 and 2 above. ●

In order to assure ourselves that the procedure in step 3 of algorithm 3.5.5 yields a well-defined labelling of the resulting tree, it is sufficient to notice that the vertex z determined at the end of step 3 has, in the original tree, the largest label among all labels of vertices smaller than the label of v. Figure 9c demonstrates deletion of vertex 10, and figure 9d shows the result of deleting vertex 11.

For the time complexity of operations we have the following simple estimate.

Theorem 3.5.6. *In a binary search tree of height h, the performance of each of the operations MEMBER, INSERT, DELETE, MIN, and MAX requires time $O(h)$.*

Proof. The performance of each of the above operations requires passing through a path which begins in a vertex of the tree (most often in the root) and is always directed downwards. In each vertex on the path we perform operations the duration of which is bounded by a constant. The length of the path cannot exceed h. ●

We now investigate the relation between the height of the binary search tree and the number of elements in the represented set. The maximum number of vertices of a binary tree of height h is $2^h - 1$ and, hence, for n elements of the set we need a tree of height $\log(n + 1)$. If we do not perform the operations *INSERT* and *DELETE*, then it is appropriate to design the tree in such a way that it has precisely that height, and then the operation *MEMBER* will require time $O(\log n)$ – see e.g. [178]. The performance of the operations *INSERT* and *DELETE* can deform the tree, and consequently the, estimate $O(n)$ for the duration of operations on a binary search tree in the worst case cannot be improved on. There are several methods of coping with that deficiency:

(a) Using more perfect, but more complicated, algorithms for execution of the operations *INSERT* and *DELETE* which will guarantee that the binary tree will always be balanced, with height $O(\log n)$. Such procedures actually do exist, and we shall present them in the following section.

(b) Relying on the non-occurrence of the deformation of the tree. The considerations presented in Chapter 10, as well as practical experience, indicate that if the choice of numbers to be stored and deleted is sufficiently random, then the height of the binary search tree will oscillate around the value $1.4 \log n$, and hence, the average length of the operations will not be much larger than the theoretical optimum. When the random character of the choice of the numbers is not guaranteed, we can improve the situation by the following randomising process: we choose a one-to-one function f defined for all the numbers to be stored, and such that $f(i)$ changes 'wildly' with i. The elements of the set are then manipulated according to the value of $f(i)$, instead of i. This, however, will render the performance of the operations MIN and MAX much more complicated.

(c) Using different tree structures in which the trees always have height $O(\log n)$, e.g. 2–3 tree [1] etc. A very detailed description of such structures can be found in [23].

3.6 Balanced binary trees

As mentioned in the preceding section, it is possible to add elements into a binary tree in such an order that the height of the tree will become substantially larger than the theoretical minimum $\log(N + 1) - 1$, where N is the number of vertices of the tree; in the worst case it can even be equal to N. This will then affect unfavourably the duration of the individual operations performed on the tree. We are going to show now how to modify the operations $INSERT$ and $DELETE$ in such a way that the height of the tree will always be $O(\log N)$.

Definition 3.6.1. An *AVL-tree*, see [42], is a binary search tree such that for each vertex v the following holds:

either v is a leaf,

or v has a single son which is a leaf,

or v has two sons of heights h_1 and h_2, satisfying $|h_1 - h_2| \leq 1$. ●

Theorem 3.6.2. *An AVL-tree of n vertices has height at most* $2 \log n$.

Proof. Denote by $N(h)$ the minimum number of vertices in an AVL-tree of height h. Obviously, $N(0) = 1$ and $N(1) = 2$. If $h \geq 2$ then the root of the tree has two sons one of which has height $h - 1$ and the other one has height either $h - 1$ or $h - 2$. The first one therefore has at least $N(h - 1)$ sons, and the other one at least $N(h - 2)$ sons, since $N(h - 2) \leq N(h - 1)$. Therefore, the whole tree has at least

$$1 + N(h - 1) + N(h - 2) \geq 2N(h - 2)$$

vertices. Consequently, $N(h) \geq 2^{h/2}$ and thus, $h \leq 2 \log N(h)$. ●

Note: by a more exact computation, the estimate in theorem 3.6.2 could be somewhat improved.

We introduce first the following auxiliary notation.

Definition 3.6.3. For each vertex v of a binary tree denote by $l(v)$ or $r(v)$ the height of the left son, or right son, respectively, of the vertex v if the son exists, else put $l(v) = -1$ or $r(v) = -1$. Further denote $b(v) = l(v) - r(v)$. The vertex v is said to be balanced if $b(v)$ takes the value $-1, 0$ or 1. ●

Thus, an AVL-tree is a binary tree in which each vertex is balanced. Let us now have a look at the mechanism of generation of possible unbalanced vertices when a vertex is added by means of algorithm 3.5.3. Such an addition can change the value $b(v)$ of an arbitrary vertex v only by decreasing it or increasing it by 1. We are thus interested in the change of $b(v)$ from 1 to 2 or from -1 to -2. Since these two cases are mirror images of each other, we shall analyse in detail the first one only.

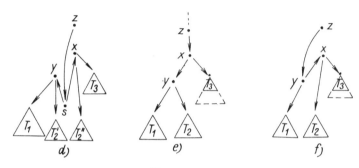

Figure 10.

Figures 10a and c demonstrate two possible ways of generating an unbalanced vertex through the operation *INSERT*. In both cases, the trees denoted by T_1, T_2 and T_3 originally had the same height (or they had all been empty) and after the new vertex was added, the height of the tree T_1 in figure 10a and of the tree T_2 in figure 10c has increased by 1, thus making the vertex x unbalanced.

The balance can be restored without affecting the balance of other vertices in the manner demonstrated in figures 10a, b and 10c, d. (In figure 10d the symbol s denotes the root of the tree T_2, while T_2' and T_2'' are the trees determined by the left and right son of s, respectively.) It can be easily seen that this leads to a binary search tree in which x as well as all of its original successors are balanced.

In order to perform the full modification, it is sufficient to change just the sons of the vertices x, y, z and, eventually, s. It is also important to notice that the height of the tree formed by vertices x and y and by the trees T_1, T_2 and T_3, which was increased during the addition of the new vertex in both cases by 1, decreases after the modification of the shape of the tree to the original value, and hence by balancing the vertex x we automatically restore the balance of all the predecessors of x, which was violated. Thus, it is sufficient to balance the vertex which is the lowest one among all unbalanced vertices.

Let us now try to characterise the vertex which has to be balanced. It is clear that the vertex x lies on the path from the root to the newly added vertex, and moreover, the following hold:

either $b(x) = 1$, when the new vertex has been added to the left of x, or $b(x) = -1$, when it has been added to the right of x.

Moreover, for each vertex y lying between x and the added vertex we necessarily have $b(y) = 0$ since otherwise one of the following would take place:

either $b(y)$ would change to 0 (from 1 or -1) or remain unchanged, but then the height of y would not be changed and consequently, no change of $b(x)$ would occur,
or $b(y)$ would be changed to 2 or -2, in which case, however, y would be balanced, making the balancing of x unnecessary.

On the other hand, it is obvious that a vertex x for which the above conditions are satisfied will have to be balanced, and those conditions determine x uniquely. With this in mind, we are able to formulate the following algorithm.

Algorithm 3.6.4. Operation *INSERT* in an AVL-tree

Input data: number C and an AVL-tree T with a labelling c of vertices and with given values $b(v)$ of all vertices v as introduced in definition 3.6.3.
Task: add to the tree T a new vertex with label C.
Auxiliary variables:

v, the vertex under consideration,

x, the vertex which is a candidate for balancing,

u and z, usually the fathers of v and x, respectively.

1. [*Beginning the search*] If the tree T is empty, we construct a tree of one vertex r and put $b(r) = 0$, $c(r) = 0$, and we terminate the computation. If T is non-empty, we set v and x equal to the root of the tree and u equal to its heading.

2. [*Traversing the tree*] Continuously repeat the following operations.

2a. If $C = c(v)$, then T contains a vertex with label C, and we terminate the computation.

2b. If $C < c(v)$ and v does not have a left son, then we add a left son w to the vertex v with label C. Then we set $b(v) := b(v) + 1$ and $b(w) := 0$, and we continue to step 3.

2c. If $C < c(v)$ and v has a left son, then the following two instructions are performed:

if $b(v) \neq 0$, set $x := v$ and $z := u$; no matter what value $b(v)$ has, set $u := v$ and $v :=$ left son of v.

2d. If $C > c(v)$ and v does not have a right son, then we add a right son w to the vertex v with label C. Then we set $b(v) := b(v) - 1$ and $b(w) := 0$, and continue to step 3.

2e. If $C > c(v)$ and v has a right son, then the following two instructions are performed:

if $b(v) \neq 0$, set $x := v$ and $z := u$; no matter what the value of $b(v)$ is, set $u := v$ and $v :=$ right son of v.

3. [*Updating the value of b*] It is necessary to change the value of b for all vertices of the tree T lying on the path from x to v (including x and v): we increase b by 1 if we proceed to the left from y in step 2c, and we decrease b by 1 if we proceed to the right from y in step 2e.

4. [*Modification of the tree*] If the value $b(x)$ after the performance of step 3 equals 2, then we have the situation illustrated by figures 10a or 10c; if the value is -2, then the situation is the mirror image of the first one. We modify the shape of the tree as indicated in figures 10a, b, c, d. For vertices to which new sons have been added we have to perform the corresponding modification of the value of b. ●

The modifications in figure 10 play a role also in fixing the unbalanced vertices resulting from the operation *DELETE* according to algorithm 3.5.5, but we have to change their interpretation: the trees T_1 and T_3 in figure 10a and the trees T_2 and T_3 in figure 10c originally had equal heights larger by 1 than the height of the tree T_2 in figure 10a or T_1 in figure 10c. By deleting a vertex of the tree T_3, the height of this tree was decreased by 1, which has in both cases made the vertex x unbalanced. We restore the balance by a modification of the tree as in figures 10b or 10d. When a vertex is deleted, the balance can also be destroyed for the reason illustrated in figure 10e, where the trees T_1, T_2 and T_3 originally had the same height, and by deleting a vertex the height of T_3 was diminished. In this case the modification indicated in figure 10f can be used. (Each of the situations in figures 10a, c, e naturally has a mirror image.)

In the cases illustrated by figures 10e, f, the height of the tree consisting of the vertices x and y and the trees T_1, T_2 and T_3 remains unchanged after the vertex has been deleted and the successive modification has been performed. On the other hand, the cases illustrated by figures 10a, b and 10c, d lead to a decrease in the height which can make some of the predecessors of x unbalanced. Therefore, the modification of the shape of the tree must in general be repeated, in contrast to the operation *INSERT*.

We can now present the detailed algorithm for deleting a vertex.

Algorithm 3.6.5. Operation *DELETE* in an AVL-tree

Input data: an AVL-tree T, a vertex v of T, and the value of b for each vertex of the tree T.

Task: delete the vertex v from the tree T, so as to get an AVL-tree.

1. [v *has at most one son*] The vertex v is deleted according to algorithm 3.5.5, step 1 or 2. Then we search the predecessors of the vertex v in the direction of the root of the tree until either the root is reached, or the computation is terminated. For each successor w we perform the following operations.

1a. If we have entered w from the left son and if $b(w) = 0$, put $b(w) := b(w) - 1$, and if the original value of b was zero, terminate the computation.

1b. If we have entered w from the right son and if $b(w) = 0$, put $b(w) := b(w) + 1$, and if the original value of b was zero, terminate the computation.

1c. When a situation occurs as illustrated either by figures 10a, c, e or its mirror image (which takes place if, and only if, neither the condition of step 1a nor that of step 1b is satisfied), perform a modification of the tree according to figures 10b, d or f or its mirror image, respectively. If the modification was performed according to figures 10e, f, or the mirror image, terminate the computation.

2. [v *has two sons*] The case that v has two sons is translated to the case in step 1 by a procedure according to step 3 of algorithm 3.5.5. ●

The operations *MEMBER*, *MIN*, and *MAX* performed according to algorithms 3.5.1 and 3.5.4 do not change the shape of the binary search tree, and consequently, they can also be applied to AVL-trees.

Following theorem 3.6.2, we can prove an estimate for the above operations.

Theorem 3.6.6. *The operations MEMBER, INSERT, DELETE, MIN and MAX performed according to algorithms 3.5.1, 3.6.4, 3.6.5 and 3.5.4 on an AVL-tree of n vertices require time $O(\log n)$.*

Proof. Using theorems 3.5.6 and 3.5.2, it is sufficient to observe that the operations *INSERT* and *DELETE* operate only on the vertices lying on the path from the root of the tree to the vertex to be inserted or deleted, or the vertices lying near. The overall number of these vertices is $O(\log n)$ and on each of them only a constant number of operations is performed. This is true even in the worst case of the operation *DELETE* when balancing of all of the vertices lying on the above path is required. ●

An important consequence follows from theorem 3.6.6: using AVL-trees, it is possible to implement a dictionary of a priority queue in such a way that each of the admissible operations is performed in time $O(\log n)$, where n is the number of elements of the dictionary or the priority queue. This fact will be used frequently throughout the book, often without an explicit reference to theorem 3.6.6.

An application of balanced trees also makes it possible to implement a mergeable heap in such a way that all the operations will only require time $O(\log n)$, see e.g. [1] or the very elegant method described in [263] which is based on a different concept of balanced trees. Since none of the main algorithms in the present book uses mergeable heaps, we shall not study their implementation any further.

Balancing a binary search tree described by algorithms 3.6.4 and 3.6.5 is a classical example of the so-called dynamisation of data structures which has recently drawn much attention, and in which a number of interesting results have been obtained, see [7, 58, 59, 199, 206, 213, 231, 239, 240, 241, 242, 260].

3.7 Heap

A heap is a very simple structure making it possible to perform rapidly the set-theoretical operations *INSERT* and *EXTRACT-MIN*.

Definition 3.7.1. A *heap* is a binary tree the vertices of which are labelled by integers, and which satisfies the following condition.

Each vertex of the heap which does not belong to the last or last-but-one layer has two sons. In the last-but-one layer there are, from left to right, first vertices with two sons, then eventually a vertex having only a left son, and finally vertices without sons.

The labelling of the vertices of the heap is performed in such a way that the label of each vertex is smaller or equal to the label of any son. ●

An important property of a heap is that its shape is uniquely determined by the number of its vertices. An example of a heap is shown in figure 11a.

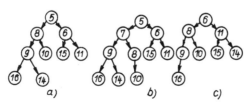

a)   b)   c)

Figure 11.

A set can be described by a heap by choosing the heap with the number of vertices equal to the number of elements of the set, and using the elements as labels of the heap. Operations are performed on heaps in the following way.

Algorithm 3.7.2. Operation *INSERT* in a heap

Input data: a heap and a number C.

Task: add a vertex labelled by C to the heap to get a new heap.

1. [*Adding a vertex*] Add a new vertex to the heap. There is only one way of performing this:

1a. If there is a vertex v in the last-but-one layer which has only a left son, add a right son to v.

1b. If 1a does not hold, and if there is a vertex in the last-but-one layer which does not have sons, add a left son to it.

1c. If each vertex in the last-but-one layer has two sons, then add a left son to the left-most vertex of the last layer (thus creating a new layer of the heap).

1d. If the heap is empty, form a single-vertex tree.

2. [*Labelling the vertex*] Label the added vertex by the number C, and set w equal to the added vertex.

3. [*Modifying the labelling*] If w is the root of the heap, the computation is terminated; else if z is the father of w, perform the following two operations:

if the labelling of z is larger than that of w, interchange their labels;

put $w := z$, and repeat step 3. ●

Algorithm 3.7.3. Operation *EXTRACT-MIN* in a heap

Input data: a heap.

Task: delete from the heap the vertex with the least label, so as to get a new heap.

1. [*Deleting the root*] We delete the root of the heap since it is the vertex with the least label.

2. [*Creating a new root*] We delete from the heap the right-most vertex of the lowest layer, insert it in place of the original root and delete w.

3. [*Modifying the labelling*] As long as w has a son with a label smaller than that of w, perform the following operation: if z is a son of w with a smaller label, interchange the labels of w and z, and set $w := z$. ●

Figure 11b demonstrates the addition of a vertex with label 7 to the heap of figure 11a, and figure 11c demonstrates the deletion of the minimum of that heap.

The estimation of speed of the above operations is analogous to other tree data structures.

Theorem 3.7.4. *Operation INSERT and EXTRACT-MIN can be performed on a heap of n vertices in time $O(\log n)$.*

Proof. A heap of height h contains at least 2^h vertices, and hence, $h \leq \log n$. The operations are performed along a path from the root to one of the leaves, and on each layer a constant number of instructions is needed. ●

The main advantage of a heap is that the strong dependence of its shape on the number of vertices makes it possible to determine sons and fathers not by means of pointers, but by a direct computation.

Theorem 3.7.5. *The vertices of a heap of n vertices can be ordered into a sequence $v_1, \ldots, v_n$ in such a way that the left and right sons of the vertex v_i are v_{2i} and v_{2i+1}, respectively, and the father of v_i is v_j for $j = \lfloor i/2 \rfloor$.*

Proof. The desired ordering is obtained by choosing as v_1 the root, then continuing with the vertices of the second layer (from left to right) followed by those of the third layer, etc. Denote by w_{ij} the jth vertex from the left of the ith layer. Since the ith layer contains 2^{i-1} vertices, the number of vertices in all the higher layers altogether, including the ith one, is $1 + 2 + 4 + \ldots + 2^{i-1} = 2^i - 1$. Consequently, $w_{ij} = v_k$ for $k = 2^{i-1} - 1 + j$. The sons of w_{ij} lie in the $(i + 1)$th layer on the positions $2(j - 1) + 1$ and $2(j - 1) + 2$ from the left, and therefore, they are the vertices v_p and v_q for

$$p = 2^{(i+1)-1} - 1 + 2(j - 1) + 1,$$

and

$$q = 2^{(i+1)-1} - 1 + 2(j - 1) + 2.$$

Therefore, $p = 2k$ and $q = 2k + 1$. This proves the statement concerning sons, and that concerning the father follows from the equations: $i = \lfloor 2i/2 \rfloor = \lfloor 2(i + 1)/2 \rfloor$. ●

Theorem 3.7.5 is illustrated by figure 12.

$$1\ 2\ 3\ 4\ 5\ 6\ 7\ 8\ 9$$

Figure 12.

It is thus possible to implement a heap profitably by means of an array $A(i)$, $i = 1, ..., n$, where $A(1)$ is the root, and the sons of $A(i)$ are $A(2i)$ and $A(2i + 1)$, whereas the father of $A(i)$ is $A(\lfloor i/2 \rfloor)$. Moreover, the majority of contemporary computers allow multiplication and integer division by 2 to be performed not as an arithmetic operation, but by shifting words by 1 bit to the left or right, respectively. The implementation of heaps makes good use of the memory of the computer, and makes it possible to perform the operations *INSERT* and *EXTRACT-MIN* quickly.

3.8 Factor set

We remind the reader of the concept of factor set. If an *n*-element set is partitioned into pairwise disjoint classes on which only the operations of factor set are performed, then each performance of *UNION* will diminish the number of classes by one, and consequently, *UNION* is performed at most $n - 1$ times. In contrast, the number of repetitions of *FIND* is unbounded.

The simplest implementation of a factor set makes use of a function F (given by an array) assigning to each element x the class $F(x)$ in which x lies. The time needed for performing $FIND(x)$ is independent of the numbers of elements in the classes of the given partition. The operation $UNION(U, V)$ is slower, since for all elements of one of the classes, for example V, we must change the value of F to $F(x) = U$.

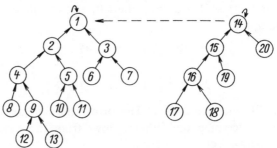

Figure 13.

A more useful implementation of a factor set uses a representation of each class of the partition by a rooted tree the vertices of which are the elements of the class to be represented. The edges of those trees will be oriented in the directions towards the root, in contrast to the usual way. For a description of such trees it is sufficient to present a function F which assigns to each vertex v its father $F(v)$. For formal reasons, we also define

$F(x) = x$ for each root x. An example of the partition of a 20-element set into the classes $\{1, ..., 13\}$ and $\{14, ..., 20\}$ is shown in figure 13 and in the first two rows of the following table:

x:	1	2	3	4	5	6	7	8	9	10	11	12	13	14	15	16	17	18	19	20
$F(x)$:	1	1	1	2	2	3	3	4	4	5	5	9	9	14	14	15	16	16	15	14
	1	1	1	2	2	3	3	4	4	5	5	9	9	1	14	15	16	16	15	14
	1	1	1	1	2	3	3	4	1	5	5	1	9							

In order to perform $FIND(x)$, we generate the sequence x, $F(x)$, $F^2(x)$, ..., until we find an element z with $z = F(z)$. This is the root of the corresponding class. For example, in figure 13 the operation $FIND(12)$ would proceed through the vertices 12, 9, 4, 2 and 1. The union of two classes with roots x and y of the representing trees is simply performed by declaring one of the roots as a son of the other one, i.e. by setting $F(x) := y$ or $F(y) := x$. This operation is shown in figure 13 by a dashed line, and in the table above in the third row.

Since the distance from the root which affects the speed of operation $FIND$ is increased by 1 for the vertices of the added tree, it is appropriate to add the smaller of the two trees to the larger one. We therefore use another array D which to each root x assigns the member $D(x)$ of vertices of the tree in which x lies. It is obvious that in the course of performing the operation $UNION$ directed by the value of D we must correct the value of D for the root of the new tree.

The above method increases the distance of vertices from the root, thus rendering slower the operation $FIND$. There is a simple trick for shortening these distances, at the cost of slowing down the operation $FIND$ only slightly.

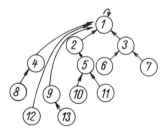

Figure 14.

For example, if we perform the operation $FIND(12)$ on the tree of figure 13, then we obtain information that all of the vertices 12, 9, 4 and 2 belong to the class represented by the root 1, and therefore, they can be directly connected to the vertex 1 in the manner indicated in figure 14 and the fourth row of the above table. This diminishes the average number of iterations of the function F when searching the elements 1 to 13, from $43/13$ to $34/13$, which is almost 20%. A disadvantage of this method is the necessity to make the passage from 12 to 1 twice because during the first passage we do not yet know where those vertices are to be connected to.

Algorithm 3.8.1. Operation *UNION* on a factor set with balancing
Task: perform the union of the sets with representatives x and y.
   If $D(x) = D(y)$, put $F(y) := x$ and $D(x) := D(x) + D(y)$; else, put $F(x) := y$ and $D(y) := D(x) + D(y)$. ●

Algorithm 3.8.2. Operation *FIND* on a factor set with shortening of paths
Task: find the representative x_0 of the set in which x lies.
   1. [*Searching for the root*] Put $x_0 := x$. Repeatedly put $x_0 := F(x_0)$ until $x_0 = F(x_0)$.
   2. [*Path compression*] Put $z := F(x)$. Perform repeatedly the following three instructions: $F(x) := x_0$; $x := z$; $z := F(x)$, until $x_0 = z$. ●

Algorithm 3.8.1 takes a constant time and, hence, all the possible $n - 1$ performances of the operations *UNION* take time $O(n)$. A precise analysis of the speed of the operation *FIND* with balancing and compressions would be very difficult. J E Hopcroft and J D Ullman first showed [1] that the performance of m repetitions of the operation *FIND* requires at most time $O(m \log^* n)$ where

$$\log^* n = \min \{i \,|\, \underbrace{\log \log \dots \log n}_{i \text{ times}} \leq 1\}.$$

The function $\log^* n$ grows to infinity, but its growth is extremely slow: for example, for all $n = 1, \dots, 2^{65\,536}$ we have $\log^* n \leq 5$. From the practical viewpoint we can thus assume that the mean time needed for one performance of the operation *FIND* is independent of the size of the set under consideration. R E Tarjan has improved this estimate using the inversion of the so-called Ackerman function, and at the same time he proved that his estimates cannot be improved for algorithm 3.8.2, see [253]. Moreover, he proved that the same is true for all algorithms satisfying a certain weak and natural condition [254]. Consequently, the mean time required for one performance of the operation *FIND* increases with the number of elements of the set under consideration, although the increase is very slow.

3.9 Representation of graphs

The simplest manner of presentation of a graph is by means of its adjacency matrix. The main advantage of this method is its simplicity which follows from the fact that the majority of programming languages make it possible to describe easily matrices and operations on them, and from the rapidity with which it can be decided whether two vertices are connected by an edge. In the case of undirected graphs we can save 50% of the memory by storing only the part of the matrix lying above the diagonal (or below), but this makes the algorithm more complicated and slows down the computation.

We often process only graphs with a small number of edges. An example is a planar graph in which the number of edges does not even reach the triple of the number of vertices. Then the application of the adjacency matrix leads to a waste of memory, and moreover, searching this matrix which mostly consists of zeros slows down the computation considerably.

The most usual method of describing a graph with a small number of edges is to assign to each vertex v a list $S(v)$ of its neighbours, implemented most often as a linked list. For the analysis of the speed of algorithms which we are going to present in Chapters 6 and 9 we shall often implicitly assume that this is precisely the method of description used. An undirected edge connecting vertices u and v occurs twice in this description: once as u in $S(v)$, and once as v in $S(u)$. In some situations (e.g. if edges are often deleted) it is suitable to add to these data both-sided references by means of pointers. Sometimes it is useful to describe explicitly the end-vertices of each edge (e.g. by arrays $END1$ and $END2$). Similar methods are also used for a description of directed graphs.

An important special case of graphs is the planar graphs. In this case we are usually also interested in the method of representation of their planar drawing. Then we can proceed with advantage as follows.

First of all, each edge is understood as a pair of directed edges with opposite directions. To each directed edge H we assign its initial vertex $INIT(H)$, its terminal vertex $TER(H)$, and the reverse edge $REV(H)$. Hence, $REV(H) = (TER(H), INIT(H))$. Furthermore, we denote by $Q(H)$ the directed edge with the same initial vertex as that of H and determined by the following property: $Q(H)$ is the first directed edge we meet when moving clockwise around the vertex $INIT(H)$ after leaving the edge H, see figure 15a. The assignment of Q is closely related to the cyclic ordering of edges of a topological planar graph mentioned already in §1.2: if $h = \{u, v\}$ is an (undirected) edge of a topological planar graph with a cyclic

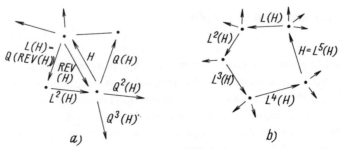

Figure 15.

ordering q of edges, and if we denote $q(u, h)$ as $\{u, w\}$, then for the directed edge $H = (u, v)$ we have $Q(H) = (u, w)$.

The assignment Q is a bijection, and hence there exists an inverse mapping Q^{-1} assigning to the directed edge H the directed edge $Q^{-1}(H)$ which is the first one we meet when moving counter-clockwise around $INIT(H)$ from H.

Further useful notation is $L(H)$ and $R(H)$ for the directed edge whose initial vertex is the terminal vertex of the directed edge H and determined as the left-most and right-most continuation of H respectively, see figure 15b. It is easy to check that for each edge H we have

$$L(H) = Q(REV(H)), \qquad R(H) = Q^{-1}(REV(H))$$

and

$$Q(H) = L(REV(H)), \qquad Q^{-1}(H) = R(REV(H)).$$

Therefore, the mappings L and Q, respectively R and Q^{-1}, need not be stored in the memory at the same time.

Using the mappings REV, Q, Q^{-1}, L and R we are able to express fairly complicated relationships and movements in planar graphs. For example, the sequence H, $L(H)$, $L(L(H))$, $L(L(L(H)))$, ... describes the rotation around the face of the topological planar graph lying to the left of H. The rotation around the face to the right is analogously expressed by means of the function R, and so on.

Exercises

1. In the language of random access machines, or in any programming language, write down programs for operations on a two-sided queue implemented as a compact list with floating ends.
2. Write down programs for operations on a doubly-linked list using a garbage collector for which a simple list is sufficient.

3. Construct an algorithm generating a binary search tree in which the operation *MEMBER* never performs a comparison of keys more than $\lceil \log(n + 1) \rceil$ times. Prove that this estimate cannot be improved.

4. In which order is it necessary to store the numbers $1, ..., n$ in a binary search tree (originally empty) in order to obtain a tree such that the performance of dictionary operations requires up to n comparisons of keys? Show that there are 2^n such permutations.

5. Write down programs for the operations *INSERT* and *EXTRACT-MIN* on a heap implemented by means of an array as described at the end of §3.8.

6. In how many ways is it possible to describe the set $\{0, 1, ..., 9\}$ by means of a heap?

7. Consider the implementation of a factor set by means of rooted trees (as described in §3.10) in such a way that the operation *UNION* uses balancing, but the operation *FIND* operates without path compression. Prove that if the number of elements of the factor set is n, then m performances of the operation *FIND* will require time $O(m \log n)$, and that this estimate cannot be improved.

8. When performing the operation *FIND* on a factor set implemented by the method described in §3.10 it is possible to shorten the path from the vertex x under consideration to the root by re-connecting each vertex from its father to its grandfather. In this way, the original path from x to the root yields two paths of half the length. This brings the advantage of an algorithm making one passage only. Write down the corresponding program for the operation *FIND* implemented in this way.

9. Describe in detail a method of representation of a graph G of n vertices and m edges, and write down a program which in time $O(n + m)$ determines the number of edges of the graph G.

4

Graph searching

When processing trees or graphs, we often need to search in a certain way all vertices of a given graph, or to perform an operation on each of them. At first sight it may seem that this problem is not too important. It turns out, however, that the opposite is true. A systematisation and a detailed analysis of the ways of searching make it possible to simplify significantly the logical structure of a number of algorithms and, hence, to make them more lucid and intelligible. Moreover, some special types of search have advantageous properties which can be used for the design of quick algorithms. Some algorithms which we shall present in Chapter 6 (see also [154, 251 and 252]) were not found until properties of graph searching were sufficiently investigated; more precisely, the idea of those algorithms was known, but by using the techniques to be described in the present chapter it has first become possible to reach a sufficient speed in the algorithms.

4.1 Searching a rooted tree

Since rooted trees are special cases of directed graphs, it would be sufficient to present searching of the latter. However, searching of trees is clearer, and it has a further significance for the following reason: we. often encounter objects with a hierarchic structure which can be represented by means of a rooted tree. As examples, let us mention arithmetic expressions and their structure, structural formulae in organic chemistry, etc. A description of such an object is, nevertheless, given by a linear sequence of symbols. All the methods of linearisation can be simply and intelligibly described by means of searching the corresponding trees.

We shall first restrict our attention to binary tree searching. We shall describe an algorithm that makes it possible to search the vertices of the tree in such a way that for the transitions we use the directed edges of the tree in both directions. At the same time we can execute operations on the vertices which are unspecified at present and which we call $OP1$, $OP2$ and $OP3$.

Algorithm 4.1.1. Depth-first searching of a binary tree
Input data: a binary tree with a root r.
Task: searching all vertices of the tree in a uniquely specified manner.
1. [*Choice of the initial vertex*] Put $v := r$.
2. [*Proceeding within the tree*] Repeat continuously steps 2a, 2b, or 2c:
   2a. [*Proceeding downwards to the left*] If the operation *OP*1 has not yet been performed on the vertex v, then we perform it and, if v has a left son w, we proceed from v to w, i.e. we put $v := w$.
   2b. [*Proceeding downwards to the right*] If the operation *OP*1 has already been performed on the vertex v, but the operation *OP*2 has not yet been performed, then we perform *OP*2 and, if v has a right son w, we proceed from v to w, i.e. we put $v := w$.
   2c. [*Return*] If the operation *OP*2 has already been performed on the vertex v, then perform *OP*3, and if v is the root of the tree, then we terminate the computation; if v is not the root, we proceed from v to its father w, and we put $v := w$. ●

We can assume that a part of the operations *OP*1 or *OP*2 is assigning the note 'we have gone left or right already'. The name of the algorithm is derived from the basic idea of proceeding to the depth of the tree as far as possible, returning only when further procedure is impossible.

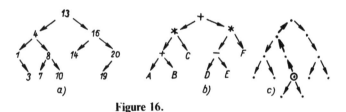

Figure 16.

The vertices of the binary tree of figure 16a will be searched by the above algorithm in the following order:

13, 4, 1, 3, 1, 4, 8, 7, 8, 10, 8, 4, 13, 16, 14, 16, 20, 19, 20, 16, 13.

If the operations *OPi* for $i = 1, 2, 3$ include printing the name of the vertex into the output file *FILEi*, then the outputs have the following form:

FILE1: 13, 4, 1, 3, 8, 7, 10, 16, 14, 20, 19
FILE2: 1, 3, 4, 7, 8, 10, 13, 14, 16, 19, 20
FILE3: 3, 1, 7, 10, 8, 4, 14, 19, 20, 16, 13.

Note that for the binary search tree in figure 16, the application of *OP*2 makes it possible to list the vertices of the tree in the order determined by their keys. Figure 16b presents the hierarchical structure of an algebraic expression with variables A, B, C, D, E and F. If the operation $OP1$ of algorithm 4.1.1 for searching the given tree consists of printing the left parenthesis, $OP2$ of printing the name of the vertex, and $OP3$ of printing the right parenthesis, then we obtain the following output:

$$((((A) + (B)) * (C)) + (((D) - (E)) * (F)))$$

which is, except for the excessive use of parentheses, the usual algebraic notation of the given expression. If, on the other hand, only the operation $OP3$ yields output which is the name of the vertex, we obtain the following output:

$$AB + C * DE - F * + ,$$

which is a listing of the same expression in the so-called reversed polish notation used in some pocket calculators and computers.

A linear listing of the vertices of a binary tree using the operation $OP1$ of algorithm 4.1.1 is called the *preorder*, the use of the operations $OP2$ and $OP3$ yields *inorder* and *postorder*, respectively.

The idea of searching makes it possible to perform easily translations of different notations of hierarchical structures. For example, the algebraic notation of the expression of figure 16b, which was constructed by inorder, can easily be translated back to the form of a tree, and a simple application of the postorder method yields a translation of the original expression to the reversed polish notation.

The whole procedure can also be modified for a general rooted tree.

Algorithm 4.1.2. Depth-first searching of a rooted tree

Input data: a rooted tree T with root r.
Task: searching the vertices of the tree T systematically.

1. [*Choice of the original vertex*] Put $v := r$.
2. [*Proceeding within the tree*] Repeat continuously the following operation:

Being in a vertex v, we choose a son w of the vertex v in which we have not yet been, and we proceed to w, i.e. we put $v := w$. If such a vertex w does not exist, then either v is the root and the computation terminates, or we return to the father of the vertex v. ●

If an ordered tree is searched, then we enter the sons of a given vertex in the given order. This rule determines the course of the search uniquely. During the search we can apply various operations to the vertices.

We shall now analyse some aspects of the performance of algorithms 4.1.1 and 4.1.2 in more detail. If the description of the searched tree uses pointers to the fathers of the vertices, then the return during the search is easy to perform. In most cases, however, the use of pointers is an unnecessary waste of memory because it is sufficient to remember the path leading from the root to the currently visited vertex. The ideal data structure for the implementation of this 'method of a thread in a labyrinth' is a stack in which we store the vertex we leave when proceeding downwards, and we remove the vertex again when returning. There is, however, a more economical way which can be described as follows: after proceeding from a vertex v to its son w, we reverse the direction of the edge $v \rightarrow w$ and hence at the time of return the edge is directed towards the father of w. After performing the return, the edge is switched again to its original direction. The situation is illustrated by figure 16c in which the currently visited vertex is denoted by a small circle.

The term search is often used in a sense wider than that illustrated by algorithms 4.1.1 and 4.1.2. It usually means a procedure such that

(a) each oriented edge is traversed precisely once,
(b) we can traverse only an edge starting either at the root, or at a vertex which has already been reached.

It is thus not prescribed that after traversing the edge (u, v) we must proceed on an edge starting at the vertex v, but we are allowed to 'jump' within the limits given by the rule (b). An example of a search in the above sense is traversing the edges of the tree of figure 16a in the following order:

13–4, 13–16, 4–1, 4–8, 16–14, 16–20, 1–3, 8–7, 8–10, 20–19.

Incidentally, if search is understood as a procedure fulfilling the above rules (a) and (b), then we have to jump also during the depth-first searching, although in this case we just jump to a direct or indirect ancestor of a vertex from which we can no longer continue. For example, the edges of the tree of figure 16a are searched in the following order:

13–1, 4–1, 1–3, 4–8, 8–7, 8–10, 13–16, 16–14, 16–20, 20–19.

The jumps are made after traversing the third, fifth, sixth and eighth edges.

The search described in the preceding paragraph proceeds very systematically in a rooted tree, passing through the successive layers of the tree. It is usually called *breadth-first searching*. We typically search first the edges whose initial vertex has been reached first. The vertices are thus waiting in a queue to be chosen to become the initial vertex of the edge under search (in contrast to the depth-first searching where they were waiting in a stack). This will be used in the formulation of the following algorithm.

Algorithm 4.1.3. Breadth-first searching of a rooted tree

Input data: a rooted tree T with root r.

Task: searching all directed edges of the tree T in a systematic manner.

Auxiliary variable: queue F containing the vertices which have already been reached and which still yield a possibility of continuation.

1. [*Choice of the initial vertex*] We place the root r into the queue F.

2. [*Searching an edge*] As long as the queue F is non-empty we repeat continuously the following operation.

Denote by v the first vertex in the queue F. Choose a directed edge h with initial vertex v which has not yet been reached, search it, and mark it as searched. If the terminal vertex of h is not yet in the queue F, put it at the end of F. If no edge h satisfies the above conditions, delete the vertex v from the queue F.   ●

Breadth-first searching is particularly useful if we want to find a leaf of the tree (or, more generally, a vertex with a certain property V) closest the root. In that case we search using algorithm 4.1.3 until we reach a leaf (or a vertex with the property V). At this instant the computation can be terminated because the desired vertex has been found.

The following theorem tells us how quick the above search is.

Theorem 4.1.4. *Each of the algorithms* 4.1.1, 4.1.2, *and* 4.1.3 *will process a rooted tree of n vertices in time* $O(n)$. (*Assuming that the times needed for the performance of additional operations, e.g. OP1 to OP3, and for the choice of the edge to be searched are constant.*)

Proof. We pass through each edge precisely once, and the number of edges is $n - 1$.   ●

4.2 Graph searching

The aim of the present section is to describe the basic method of searching undirected graphs. We shall mention the directed case at the end. We are again going to proceed systematically in such a way that each edge will be traversed precisely once, and we always start at a vertex already reached. Exceptions to the last rule are at the beginning of the search and, in the case of disconnected graphs, when a transition to a new component is made. The following is the basic scheme.

Algorithm 4.2.1. Searching a graph
Input data: an undirected graph **G**.
Task: searching vertices and edges of the graph **G** in a systematic manner.
Auxiliary variables:

 U, the set of all edges which have not yet been searched,

 N, the set of all vertices which have not yet been searched,

 D, the set of all vertices which have been searched, but which have outcoming edges which have not yet been searched.

 1. [*Initiation*] Put $N :=$ the set of all vertices of the graph **G**,
$U :=$ the set of all edges of **G**, and $D = \emptyset$.

 2. [*The termination test*] (At this step we always have $D = \emptyset$.) If $N = \emptyset$
the computation is terminated.

 3. [*The choice of the first vertex in a component*] Choose a vertex v in N,
delete v from N and put $D := \{v\}$.

 4. [*The choice of the initial vertex of the edge to be searched*] Choose an
arbitrary $w \in D$.

 5. [*The test whether w is usable*] If there exists no outcoming edge
of the vertex w which lies in U, then delete w from the set D, and if afterwards D is empty, return to step 2, else to step 4. If such an edge does exist,
continue to step 6.

 6. [*Traversing an edge*] Choose an outcoming edge $h = \{w, z\}$ of the
vertex w lying in U. Traverse the edge h in the direction from w to z,
delete it from U, and if $z \in N$ delete z from N and add it to D. Return
to step 4. ●

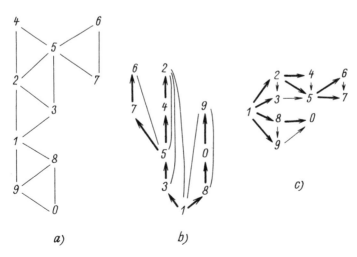

a)   b)   c)

Figure 17.

As an illustration we present three examples of searching the graph in figure 17a. [Instead of $\{x, y\}$ or (x, y) we write just xy.]

$$12, \ 18, \ 19, \ 25, \ 53, \ 89, \ 80, \ 56, \ 67, \ 75, \ 09, \ 13, \ 24, \ 54, \ 23;$$
$$13, \ 35, \ 57, \ 76, \ 65, \ 54, \ 42, \ 21, \ 23, \ 25, \ 18, \ 80, \ 09, \ 91, \ 98;$$
$$12, \ 13, \ 18, \ 19, \ 23, \ 24, \ 25, \ 35, \ 80, \ 89, \ 90, \ 45, \ 56, \ 57, \ 67.$$

Now we eva'uate the speed of this search.

Theorem 4.2.2. *Assuming that the choice of an edge and the passage through it take constant time, then algorithm 4.2.1 will search a graph of n vertices and m edges in time $O(n + m)$.*

Proof. Each performance of one of the steps 1 to 6 above takes a constant time. The sum of the numbers of elements of the sets N, D and U at the beginning is $n + m$ and it can never grow during the computation. The computation is performed in such a way that after the sequence of steps 1, 2, 3, 4, the following cycles are continuously repeated: 5, 6, 4, 5 or 5, 4, 5 or 5, 2, 3, 4, 5. In the first cycle the sum of the numbers of elements of the sets N, D and U is decreased by one at step 6, in the other two cycles it is decreased already at step 5. Consequently, the number of repetitions of the above cycles is at most $n + m$ since the computation terminates when $N = \emptyset$. ●

The following theorem shows the basic properties of the above search.

Theorem 4.2.3. *Algorithm 4.2.1 traverses each edge of the graph precisely once, and it searches the graph according to components. That is, after the choice of vertex in step 3, the steps 4, 5 and 6 are performed until all edges and all vertices of the component containing the chosen vertex have been traversed.*

Proof. The role of the set U prevents a repeated passing through an edge. Assume now that a vertex v has been chosen in step 3, and that steps 4, 5 and 6 follow. Denote by t the time instant in which a jump to step 2 is performed.

Let $v = v_0, ..., v_k$ be a path in the graph G. The vertex v_0 has been inserted into D before the instant t. Let i be the largest number such that the vertices $v_0, ..., v_i$ have been inserted into D before t. If $i < k$ then, since at time t we have $D = \emptyset$, all edges incident with v_i have been traversed, which particularly applies to $\{v_i, v_{i+1}\}$. However, if the last edge had been traversed in the direction from v_{i+1}, we would have had $v_{i+1} \in D$ previously, and if it had been traversed in the opposite direction, we would have put v_{i+1} into D in the next step, which means before t. Since both of these cases lead to a contradiction, we conclude that $i = k$. Therefore, all vertices $v_0, \ldots, v_k$ have been put into D before t and they were subsequently deleted from D, which implies that all the edges of the path $v_0, \ldots, v_k$ have been traversed before t.

We conclude that the whole component containing the vertex v has been searched before t. ●

As in the case of rooted trees, the most important methods of search in the present section are also depth-first and breadth-first searching. Again, depth-first searching is based on the vertex reached last, whereas the breadth-first procedure traverses the vertices systematically in the order in which they have been reached.

Algorithm 4.2.4. Depth-first searching a graph
We proceed as in algorithm 4.2.1, implementing D as a stack, that is, in step 4 we always choose the vertex which is in D for the shortest time. ●

Algorithm 4.2.5. Breadth-first searching a graph
We proceed as in algorithm 4.2.1, implementing D as a queue, that is, in step 4 we always choose the vertex which is in D for the longest time. ●

The ordering in which we pass through the vertices of the graph when using algorithm 4.2.4 or 4.2.5 can be completely determined by presenting an ordering of the vertices of the graph under search and then, in step 3, always choosing as v the first vertex of the set N, and if in step 6 there is a choice between edges $wz_1, \ldots, wz_k$, then choosing that for which z_i is the least element in the presented ordering.

When searching a connected, topological, planar graph, the course of the depth-first or breadth-first searching can be completely determined by choosing the first edge, directing it, and applying the rule that we always proceed to the left-most edge.

Algorithm 4.2.6. Searching a topological planar graph
Input data: a topological planar graph G with a cyclic ordering of edges q, and a directed edge $h = (x, y)$ of the graph G.
Task: search G.
Auxiliary variables: for each vertex z choose an outcoming directed edge h_z.
Proceed as in 4.2.4 or 4.2.5 with the following specifications. In step 3 choose the vertex x, and let h be the first edge to be traversed. For the vertex x, choose h as h_x, and for each other vertex w let h_w be the reversely directed edge to that through which w has been passed for the first time.

The choice of edge in step 6 is such that the outcoming edges of the vertex w are traversed in clockwise order, starting with h_w. ●

Traversing the edge $\{w, z\}$ in step 6 of algorithm 4.2.1 directs this edge in the direction from w to z. By theorem 4.2.3 we see that each search determines an orientation of the given graph. Figures 17b and 17c show orientations of the graph of figure 17a determined by depth-first and breadth-first searching, respectively. Those directed edges (w, z) through which we entered the edge z in step 6 which had not been reached previously (i.e. had not been an element of N) will be called *basic*. The basic directed edges of each search give rise to a rooted tree in each component of the graph. (Prove it!) The root of this tree is that vertex of the component which was chosen in step 3 as the initial edge of the component. We shall call this collection of trees the *forest of search*, and in the case of a connected graph, the *tree of search*. Trees of search are marked in figures 17b, c by bold lines.

Trees of the depth-first and breadth-first searching have some useful properties.

Theorem 4.2.7. *If* $\{w, z\}$ *is an edge of a connected graph* **G**, *then in an arbitrary tree of depth-first searching of the graph* **G**, *one of the vertices w and z is a successor, direct or indirect, of the other one.*

Proof. If in the course of the search we have stored the vertices $v_0, \ldots, v_k$ into the stack in their present order, then the directed edges $(v_0, v_1), \ldots, (v_{k+1}, v_k)$ belong to the tree of search. Assume we have first entered w. From the moment of storing w in D until that of deleting w, all the vertices reached in that period are direct or indirect successors of w. When w is being deleted from D, the edge $\{w, z\}$ must have already been traversed, and hence, z has been reached in that period. ●

The assertion of the above theorem is illustrated by figure 17b. We are going to use it in § 6.5 for searching articulations of a graph. Another property is typical for breadth-first searching.

Theorem 4.2.8. *The shortest path from a vertex* v *to a vertex* w *of a connected graph* **G** *is given by the directed path from* v *to* w *in the tree of breadth-first searching of the graph* **G** *with the initial vertex* v.

Proof. The breadth-first search starting at v first passes through all the neighbours of v, then all the neighbours of the neighbours, etc. This implies the statement of the theorem. ●

From the above theorem and from theorem 4.2.2 it follows that the shortest path connecting two vertices of a graph **G** can be found in time proportional to the number of vertices and edges. The assertion of theorem 4.2.8 is illustrated by figure 17c.

For searching a directed graph we can just modify 4.2.1 as follows.

Algorithm 4.2.9. Searching a directed graph
Input data: a directed graph G.
Task: search the vertices and edges of G in a systematic manner.
Proceed as in 4.2.1 with the following modifications:

U is the set of all the directed edges which have not yet been searched.

D is the set of all vertices already reached which are the initial vertices of the directed edges not yet searched.

In step 1 put $U :=$ the set of all directed edges of G.

In step 5 delete w from D if it is not the initial vertex of any directed edge lying in U.

In step 6 choose a directed edge $h = (w, z)$ having the initial vertex w and lying in U. ●

The implementation of D as a stack or a queue leads to the breadth-first or depth-first search of the directed graph respectively. It is not hard to prove the analogy of theorem 4.2.2, but, on the other hand, analogies of theorems 4.2.3 and 4.2.7 need not be true. The depth-first search of a directed graph will be used in § 6.5 in order to find strong components of a directed graph.

4.3 Backtracking and the method of branches and bounds

This section is devoted to techniques which are often used for the design of combinatorial algorithms. For their presentation we shall use the search of rooted trees.

The general type of tasks which we are going to solve can be formulated as follows.

Let $X_1, ..., X_n$ be given sets (not necessarily pairwise distinct), let $A \subset X_1 \times ... \times X_n$ be a set, the elements of which are called *admissible solutions*, and let F be a *goal function* which assigns to each admissible solution a real number. The task is to find an admissible solution $(x_1, ..., x_n) \in A$ such that the value $F(x_1, ..., x_n)$ is minimal.

The solution will be searched in the form of a sequence of partial solutions. This means a sequence $(x_1), (x_1, x_2), ..., (x_1, ..., x_n)$ of vectors which will yield the required solution by a successive addition of coordinates. It is reasonable to consider just those partial solutions which can be extended to an admissible solution. However, it is often difficult to recognise which partial solutions have this property. Therefore, we are going to consider all partial solutions for which the existence of an extension is not excluded at first sight. In general, we are going to assume that sets $A_1, \ldots, A_n = A$ are given, the elements of which are called *admissible partial solutions*, in such a way that the following hold:

$$A_i \subset X_1 \times \ldots \times X_i$$

and

$$(x_1, \ldots, x_i, x_{i+1}) \in A_{i+1} \Rightarrow (x_1, \ldots, x_i) \in A_i.$$

The latter condition means that each admissible partial solution of $i + 1$ coordinates can be obtained as an extension of a partial solution of i coordinates. For technical reasons it is suitable to add another set A_0 containing the empty sequence alone which will also be considered as a partial solution, and will be denoted by Λ.

The following rooted tree will be called the *problem tree*: vertices are all admissible partial solutions with the root in the empty solution and the sons of $(x_1, \ldots, x_i)$ are just all of its extensions $(x_1, \ldots, x_{i+1})$ in A_{i+1}. The leaves of the tree are all admissible solutions lying in A, and all partial solutions which cannot be extended.

The basic variant of the method called backtracking is nothing more than depth-first searching of the problem tree while computing systematically all the successive partial admissible solutions and choosing successively the solution with the lowest value of F. The problem tree is usually immense, and therefore we try to determine partial solutions which will not be extended as 'non-prospective'. In this way we can often reduce the computation to a fragment of the original length.

We usually employ a lower estimate L and an upper estimate U of the value of the goal function F which have the property that each admissible solution fulfils $L \leq F(x_1, \ldots, x_n)$, and there exists an admissible solution fulfilling $F(x_1, \ldots, x_n) \leq U$. Sometimes such numbers are known in advance, sometimes we choose them as $-\infty$ and $+\infty$, respectively, and then we specify them in the course of computation. Usually we can, for each admissible partial solution, determine a lower bound of the values of the goal function of its extensions. We can thus assume that we are given functions F_i defined on the sets A_i of partial solutions which fulfil

$$F_1(x_1) \leq F_2(x_1, x_2) \leq \ldots \leq F_n(x_1, \ldots, x_n) = F(x_1, \ldots, x_n).$$

The following information can be derived from the facts above.

(a) If an admissible solution $(x_1, ..., x_n)$ has been found, then if $F(x_1, ..., x_n) = L$ we can terminate the computation because the solution is optimal, and if $F(x_1, ..., x_n) < U$, we can make the upper bound sharper.

(b) If a partial admissible solution $(x_1, ..., x_n)$ has been found which fulfils $F_i(x_1, ..., x_i) > U$ then it is useless to try extending it.

(c) If $F_i(x_1, ..., x_i) = U$, and an admissible solution with the value U of the goal function is known already, then, again, we are not going to try extending the partial solution.

It sometimes happens that, no matter how ingenious our choice of A_i and F_i is, we are not able to make the computation fast enough to conclude it successfully. In such a case we try to find a good approximation to the solution.

In the first place, we can diminish the sets of admissible partial solutions, even paying the price of deleting a partial solution which, although it does not look hopeful, would actually lead to the minimal admissible solution.

Another possibility is to increase the functions F_i in such a way that we no longer have, in general, the inequality $F_i(x_1, ..., x_i) \leq F_{i+1}(x_1, ..., x_i, x_{i+1})$ and, hence, it can happen again that because of an unfavourable estimate of F_i we shall delete a partial solution which would lead to a minimal solution.

It is sometimes difficult to choose the sets A_i and the functions F_i in a way that would make the computation sufficiently quick, and yet would give an approximate solution sufficiently close to the minimal solution. A great deal of feeling for the problem under consideration is required.

An application of the above methods for getting an exact or an approximate solution will be discussed in Chapter 9, where several algorithms will be presented.

Exercises

1. Write down a program for searching a rooted tree using the reversion of arrows described at the end of §4.1.
2. Prove that orientation of an undirected tree in the direction from a chosen vertex can be determined, using search, in time $O(n)$, where n is the number of vertices of the tree.
3. Given sets A and B, put

$$A \div B = (A \cup B) - (A \cap B).$$

Prove that $\div$ is a commutative and associative operation. If $C, C_1, ..., C_k$

are cycles in a graph, we say that the set $C_1, ..., C_k$ of cycles
generates the cycle C if $H \div H_1 \div H_2 \div ... \div H_k$, where $H, H_1, ... H_k$
are the sets of all edges of the respective cycles. We say that
$B = C_1, ..., C_k$ is a cyclic base of the graph G provided that the
following hold:
    (a) each cycle of G is generated by a subset of the set B,
    (b) no proper subset of B has the property (a).
Explain how a cyclic base of the graph can be found by means of
search. (*Hint*: each edge which does not belong to the tree of search
of G determines, together with the tree, a cycle from the given
cyclic base.)

4. Let G_1 and G_2 be topological planar graphs of n vertices, and let
$g_i = (x_i, y_i)$ be directed edges of G_i for $i = 1, 2$. Find an algorithm which
in time $O(n)$ either presents an isomorphism $f: G_1 \to G_2$ which fulfils
$f(x_i) = y_i$ for $i = 1, 2$ and is compatible with the planar drawing, or
proves the non-existence of such an f. (See definition 6.8.1.)

5. Describe a procedure which in time $O(n)$ determines all faces of
a topological planar graph of n vertices and shows which edges are
incident with the individual faces.

6. Using the methods of §4.3, write down a program which finds a path
of a knight on a chessboard in such a way that each square is traversed
exactly once.

5

Sorting

In this chapter we shall study methods for sorting, i.e. ordering given items $Z_1, ..., Z_n$ with respective keys $K_1, ..., K_n$ in such a way that the key values are subsequently non-decreasing (or non-increasing). Thus, the task is to find a permutation $i_1, ..., i_n$ of the numbers $1, ..., n$ such that

$$K_{i_1} \leqq K_{i_2} \leqq ... \leqq K_{i_n} \quad (\text{or } K_{i_1} \geqq K_{i_2} \geqq ... \geqq K_{i_n}).$$

This problem quite often appears in commercial data processing, but it also plays an important role in the design of combinatorial algorithms.

The data and sequences to be processed are often very large, and sorting tens of thousands of items is not exceptional. It is, therefore, necessary to find methods which would solve the problem of sorting very quickly. The importance of the problem has initiated a considerable interest in it, and a number of really quick algorithms have been found. The extent of our book does not permit us to present the methods of sorting in their full range. (For example, D E Knuth's book [23], which every reader more deeply involved in sorting should read, devotes 379 pages to this problem.) This, however, is also unnecessary because the algorithms for sorting will be needed only as auxiliary procedures for the design of combinatorial algorithms in subsequent chapters. It is our primary goal to show that sorting n items can be performed in time $O(n \log n)$ with memory $O(n)$, and in some special cases, even in time $O(n)$.

We shall usually deal only with algorithms yielding a non-decreasing ordering since the non-increasing one can be obtained by small modifications of the procedure.

5.1 Sorting in time $O(n \log n)$

There are a number of methods for sorting n numbers in time $O(n \log n)$, see [1, 23 and 140]. In our book we restrict ourselves to one method called Heapsort. Our choice was influenced by the fact that, using the description of the implementation of heaps presented in § 3.7, it is possible to explain the principle of Heapsort very concisely.

Algorithm 5.1.1. Heapsort

Input data: items $Z_1, ..., Z_n$ with keys $K_1, ..., K_n$.
Task: sort the input items to a sequence with non-decreasing values of keys.
Auxiliary variable: heap p, the vertices of which are the input items.

1. [*Fill the heap*] We fill the originally empty heap using the operation $= INSERT$ until all of the input items are vertices, in an arbitrary order.

2. [*Delete*] As long as the heap is non-empty, we delete items from it using the operation $EXTRACT-MIN$, and we store them successively into the output sequence. ●

Theorem 5.1.2. *The algorithm* 5.1.1 *requires time* $O(n \log n)$.

Proof. By theorem 3.7.4, one performance of the operation $= INSERT$ or $EXTRACT-MIN$ requires time $O(\log n)$. Each of these operations has to be performed n times. ●

It is advantageous to use the implementation of heap by means of an array as described at the end of § 3.7. We use an array of n elements, the first d of which are going to be used for a description of the heap (where d is the number of items in the heap), whereas the remaining $n - d$ will first be used for the remainder of the initial sequence of items, and then for the output sequence to be created. Since the elements are going to be stored from left to right into the output sequence, the resulting sequence will be non-increasing. If the non-decreasing ordering is required, we change the implementation of the heap in such a way that it allows a quick performance of $EXTRACT-MAX$ instead of $EXTRACT-MIN$. This variant of the algorithm is the original Heapsort, see [270].

A procedure of sorting using the algorithm Heapsort is illustrated by the example in the following table in which the heap is always stored to the left of the division dot.

Step 2	Step 3
5 2 7 4 1 6	1 2 6 5 4 7.
5. 2 7 4 1 6	2 4 6 5 7. 1
2 5. 7 4 1 6	4 5 6 7. 2 1
2 5 7. 4 1 6	5 6 7. 4 2 1
2 4 5 7. 1 6	6 7. 5 4 2 1
1 2 7 5 4. 6	7. 6 5 4 2 1
1 2 6 5 4 7	7 6 5 4 2 1

The reader is advised to check the performance of the operations $= INSERT$ and $EXTRACT\text{-}MIN$ of algorithms 3.7.2 and 3.7.3.

5.2 Quicksort

The method of sorting we are now going to describe is a variant of the algorithm designed by C A R Hoare [149] under the name Quicksort. The idea is simple: we choose a number k and rearrange the sequence of items $Z_1, ..., Z_n$ in such a way that the items with key values smaller than k are at the beginning, then follow those with key values equal to k, and at the end there are the items with key values larger than k. We further sort the first and third parts of the sequence, thus sorting the whole sequence.

But how are we to sort the first and third parts? We just repeat the same procedure. We first suitably store the third part, then we choose a number k_1 and divide the first part into three sub-parts which are stored (by the same method), thus sorting the first part of the original sequence. Then we recall the third part and process it in the same manner.

The subdivision does not continue infinitely long because soon some part of the section being processed is empty or a singleton, and then it is not processed any longer. Naturally, it is important to choose well the numbers according to which the keys of the items are sorted. We are later going to mention some methods of choice.

An example of sorting in this manner is presented by the following table in which the parts of the sequence which have not yet been sorted are denoted by parentheses and brackets, the latter denoting the parts to be sorted next. The right-hand column shows the number k used for the subdivision.

```
[8 27 15 17  2 12 23  25  9  7 29  5 15  1   15 20 16]        15
[8  2 12  9  7  5  1] 15 15 15 (27 17  23 25   29 20 16)        7
[2  5  1]  7 (8 12  9) 15 15 15 (27 17  23 25   29 20 16)        5
[2  1]  5  7 (8 12  9) 15 15 15 (27 17  23 25   29 20 16)        2
 1  2   5  7 [8 12  9] 15 15 15 (27 17  23 25   29 20 16)        9
 1  2   5  7  8  9 12  15 15 15 [27 17  23 25   29 20 16]       24
 1  2   5  7  8  9 12  15 15 15 [17 23  20 16] (27 25 29)       20
 1  2   5  7  8  9 12  15 15 15 [17 16] 20 23  (27 25 29)       17
 1  2   5  7  8  9 12  15 15 15  16 17  20 23  [27 25 29]       27
 1  2   5  7  8  9 12  15 15 15  16 17  20 23   25 27 29
```

In the following algorithm, we use a stack as an auxiliary data structure into which information is stored about the parts which are yet to be sorted.

Algorithm 5.2.1. Quicksort

Input data: items $Z_1, ..., Z_n$ with keys $K_1, ..., K_n$.

Task: sort the items.

Auxiliary variables: a stack S the elements of which are pairs of integers from 1 to n (the bounds of the parts yet to be stored), and natural numbers L and R determining the bounds of the part currently sorted.

1. [*Initiation*] Put $L := 1$ and $R := n$; S is empty.
2. [*Choice of the division point*] Choose a number k.
3. [*Division of the part to be sorted*] The items $Z_L, ..., Z_R$ are rearranged and subdivided into three parts as follows

$$Z_L, ..., Z_{J1}; \quad Z_{J1+1}, ..., Z_{J2-1}; \quad Z_{J2}, ..., Z_R$$

in such a way that the key of the item Z_i is

smaller than k for all $i = L, ..., J1$,
equal to k for all $i = J1 + 1, ..., J2 - 1$,
larger than k for all $i = J2, ..., R$.

Some of these parts are allowed to be empty.

4. [*Choice of the next part to be stored*] If $L < J1$, perform the following two instructions:

If $J2 < R$, store $(J2, R)$ into the stack S.
Put $R := J1$ and return to step 2.

If $L \geq J1$, perform the following instructions:

If $J2 < R$, put $L := J2$ and return to step 2.
If $J2 \geq R$ and the stack S is non-empty, delete a pair (X, Y) from S, put $L := X$ and $R := Y$, and return to step 2.

5. [*Termination*] The sequence is sorted. ●

Now we are going to present a variant of algorithm 5.2.1 which, for the sake of speeding up the computation, subdivides the section of the sequence being processed into two parts only. In the left one there are all the items with key value smaller than the given number k, and some of those with key value k, and the rest are in the right one. The subdivision of the section together with the rearrangement of the items is performed by means of two variables BL and BR, 'runners' which at the beginning have values $BL = L$ and $BR = R$, and which move along the section in such a way that to the left of BP there are items which have key values at most k, and hence belong to the left part, whereas to the right of BR there can lie only items with key value at least k. The movement of the runners is performed as follows:

if the key of Z_{BL} is at most k, then BL can be moved to the right,
if the key of Z_{BR} is at least k, then BR can be moved to the left.

These operations are performed until either $BR < BL$, which means that the subdivision terminates, or $K_{BL} > k > K_{BR}$, in which case the values of Z_{BL} and Z_{BR} can be interchanged, and the whole procedure repeated.

Algorithm 5.2.2. Modification of algorithm 5.2.1

The input data, task and auxiliary variables are the same as in 5.2.1, with two variables BL and BR added.

1. [*Initiation*] Put $L := 1$ and $R := n$; S is empty.
2. [*Choice of the division point*] Choose a number k.
3a. [*Initiation of the runners*] Put $BL := L$ and $BR := R$.
3b. [*The shift of BL*] While $BL \leq BR$, and the key of the item Z_{BL} has value smaller than or equal to k, increase BL by 1.
3c. [*The shift of BR*] While $BL \leq BR$, and the key of the item Z_{BP} has value larger than or equal to k, decrease BR by 1.
3d. [*Interchange of items*] If $BL > BR$, go to step 4, else interchange the items Z_{BL} and Z_{BR} and put $BL := BL + 1$, $BR := BR - 1$ and go to step 3.
4. [*Choice of the next part to be sorted*] If $L < BR$, perform the following three instructions:

If $BL < R$, store (BL, R) into the stack S, put $R := BR$ and go to step 2.
If $L \geq BR$ then if $BL < R$ put $L := BL$ and go to step 2;
If $BL \geq R$ and the stack S is non-empty, delete a pair (X, Y) from S, put $L := X$ and $R := Y$, and then go to step 2.
5. [*Termination*] The sequence is sorted. ●

We are now going to have a look at the speed of the above algorithms. There exist constants c and d such that between two successive entries in step 2 (which is the time needed for subdividing the given section into two or three parts) the number of performances of the operations is at most $c(R - L) + d$ where R and L are the bounds of the section being processed. Therefore, if the time needed for sorting n items is denoted by $T(n)$, then the following hold:

$$T(0) = T(1) = 0$$

and

$$T(n) \leq cn + d + T(n_1) + T(n_2),$$

where n_1 and n_2 are the lengths of the parts of the current section, which are obtained in the first subdivision using the number k, and which are to be sorted

In the case when the obtained parts are approximately equally large, we can assume that n_1 and n_2 are both approximately equal to $n/2$, and we obtain the following recursive formula

$$T(n) \leqq cn + d + 2T(n/2),$$

which implies

$$T(n) \leqq cn \log n + 2nd.$$

As will be shown in § 10.2, such a subdivision often takes place and then the mean time of computation is $O(n \log n)$.

If we disregard the possibility of choosing the number k so badly that all the key values of the items being processed are larger than k, or all are smaller than k, the next extreme case is $n_1 = n - 1$ or $n_2 = n - 1$. Then $T(n) \sim cn + d + T(n - 1)$, and hence, $T(n) \sim cn(n + 1)/2 - c + d(n - 1) = \Omega(n^2)$.

From the above consideration it is obvious that the speed of algorithms 5.1.1 and 5.1.2 depends heavily on the choice of the number k in step 2. We usually choose k from the key values of the items to be sorted. The choice $k = K_R$ or $k = K_L$ is not quite suitable, as will be seen in the exercises, because this leads to a long time for processing almost-sorted sequences in cases often met in practice. Good results can be obtained by choosing the key of the item lying around the centre of the section to be subdivided, i.e. $k = K_i$ for $i \doteq (R + L)/2$. Although in this case the worst time of computation is also proportional to n^2, the mean time of computation, both theoretically and in practical situations, is very short. There are also more complicated methods for choosing k, for example as the arithmetic mean of the keys of items lying in the first, second and third quarter of the section to be subdivided, etc.

5.3 Lower bounds

In this section we are going to prove that for a certain class of algorithms for sorting, including those described in §§ 5.1 and 5.2, the estimate $O(n \log n)$ of the time needed for sorting the worst-case sequence of n items cannot be improved. The class consists, roughly speaking, of all algorithms which control the computation by means of comparing keys of two items, or a key and a given number.

In order to present the precise definition of the class of algorithms, we are going to assume that each algorithm under consideration is described as a program for a random access machine, and the task of the algorithm is to sort the sequence of numbers on the input tape.

We introduce the following auxiliary concept.

Definition 5.3.1. By a description of computation of a random access machine is meant a sequence $(p_1, i_1), (p_2, i_2), \ldots$ of pairs of numbers, where p_j and i_j are the contents of the program register and the index register respectively, after the performance of the jth step of the computation under consideration. ●

The description of computation determines uniquely which instruction will be performed in each step of the computation and on which memory registers it will operate.

Now, let $(p_1, i_1), (p_2, i_2), \ldots$ and $(p'_1, i'_1), (p'_2, i'_2), \ldots$ be two descriptions of computation of the same algorithm, and let j be the least number with $(p_j, i_j) \neq (p'_j, i'_j)$. Then one of the following two possibilities occurs. Either the jth step of computation is a conditional jump instruction, and the jump was executed in one case while in the other it was not (in which situation we have $i_j = i'_j$), or the jth step is the instruction $STORE$ changing the number stored in the index register to different numbers for the two computations (and then clearly $p_j = p'_j$). In the latter possibility the goal usually is to permit the next instruction of memory move or arithmetic operation using indexed addressing to operate on different memory registers for the two computations.

Controlling the computation by means of conditional jump instructions is typical for the algorithms presented in §§ 5.1 and 5.2, whereas the algorithms which will be introduced in § 5.4 control the computation by means of addresses stored in the index register.

Definition 5.3.2. A sorting algorithm (described as a program for a random access machine) is called *comparison sorting* provided that for each sequence $a_1, \ldots, a_n$ of numbers and arbitrary permutations $r_1, \ldots, r_n$ and $s_1, \ldots, s_n$ of the numbers $1, \ldots, n$ the following holds: let $(p_1, i_1), (p_2, i_2), \ldots,$ and $(p'_1, i'_1), (p'_2, i'_2), \ldots$ be descriptions of computations of the considered algorithm sorting the sequences $a_{r_1}, \ldots, a_{r_n}$ and $a_{s_1}, \ldots, a_{s_n}$ respectively. Then the least number k with $(p_k, i_k) \neq (p'_k, i'_k)$ fulfils $p_k \neq p'_k$. ●

It is our aim to prove the following theorem.

Theorem 5.3.3. *For each comparison sorting algorithm and each natural number n there exists a permutation of the numbers* $1, ..., n$ *the sorting of which requires time at least* $\log n!$.

Proof. Let each permutation of the numbers $1, ..., n$ be processed in time at most t. To each of these permutations Π we assign the sequence $P(\Pi)$ of 0s and 1s with 1 in the jth position if, and only if, the jth step of computation for sorting the permutation Π is a conditional jump instruction which is actually executed.

The descriptions of computation for processing two distinct permutations Π_1 and Π_2 must be different. Since we are considering comparison sorting, the sequences $P(\Pi_1)$ and $P(\Pi_2)$ must also be different. The number of such sequences is at most 2^t and, at the same time, their number is equal to $n!$. Consequently, $2^t \geq n!$. ●

Since $\log n! \geq cn \log n$ for a suitable constant c, the estimate $O(n \log n)$ for sorting n items by a comparison sorting algorithm is the best possible. It can be shown that algorithm 5.2.1 is comparison sorting; it is, up to a multiplicative factor, an optimal comparison sorting algorithm from the point of view of the worst-case behaviour.

5.4 Sorting by distribution

The sorting algorithms described in §§ 5.1 and 5.2 can be universally applied, but their speed is bounded by $\Omega(n \log n)$ as we have seen in § 5.3. We are now going to present a class of algorithms which are applicable to some cases only, but whenever they are, the sorting time for n items is $O(n)$.

The basic variant of this procedure can be applied in the case where the keys of the items are integers lying between two numbers L and U which are not too different in magnitude. It is usually required that the difference $U - L$ does not exceed the order of thousands, and the smaller it is, the more suitable sorting by distribution becomes.

The method under consideration proceeds as follows: we create 'shelves' denoted by numbers $L, L + 1, ..., U - 1, U$, which, at the beginning of the computation, are empty, and then we search the items successively, storing the item with key K to the Kth shelf. Finally we order the contents of the shelves successively.

This procedure was commonly used by punch-card machines, where the shelves were physical, holding the cards. For an electronic computer it is suitable to represent the 'shelves' by lists. When applying linked lists, it is suitable to use an array $HEAD(i)$, $i = L, ..., U$, containing the headings of the individual lists, and the item with key K is stored in the list with heading $HEAD(K)$. The application of compact lists meets the following obstacle: it is not clear in advance what number of items will be stored in the individual shelves, and to plan each shelf for all of the n items would not be economical. The following algorithm solves this problem by counting first, for each $K = L, ..., U$, the number of items with key K, and then dividing the assigned memory region into shelves.

Algorithm 5.4.1. One-passage sorting by distribution
Input data: integers L and U and items $Z_1, ..., Z_n$ with keys $K_1, ..., K_n$ the values of which are integers between L and U inclusive.
Task: sort the items according to their key values.
Auxiliary variables: output sequence $W_1, ..., W_n$ in which the input items are going to be sorted, and integers C_i for $i = L, ..., U$.

1. [*Initiation*] For $j := L, ..., U$ put $C_j := 0$.
2. [*Specification of the extent of the shelves*] For $i = 1, ..., n$ put $K := K_i$ and $C_K := C_K + 1$ (after execution, C_K is equal to the number of items with key K).
3. [*Specification of the shelves*] For $j = L + 1, ..., U$ put successively $C_j := C_{j-1} + C_j$. (After execution, C_j is the number of items with key at most j.)
4. [*Distribution*] For $i = n, n - 1, ..., 2, 1$ execute successively the following instructions:

$$K := K_i, \qquad C := C_K, \qquad W_C := Z_i \quad \text{and} \quad C_K := C_{K-1}. \quad \bullet$$

We have chosen the reverse ordering of items in step 4 in order to guarantee the following property of the algorithm.

Definition 5.4.2. A sorting algorithm is said to be stable if the order of two arbitrary items with the same key in the output sequence is the same as that in the input sequence. $\bullet$

For example, given items Z_1 to Z_6 with keys $1, 3, 2, 5, 3, 6$, then the algorithm yielding the output sequence $Z_1, Z_3, Z_5, Z_2, Z_4, Z_6$ is not stable — in fact, the only stable sorting is the following:

$$Z_1, Z_3, Z_2, Z_5, Z_4, Z_6.$$

Theorem 5.4.3. *Algorithm* 5.4.1 *is stable.*
Proof. This is obvious, and we leave it to the reader as an exercise. $\bullet$

We are now going to generalise algorithm 5.4.1 to the case when each of the items $Z_1, ..., Z_n$ has not just one key, but several keys. We assume that the number of keys is fixed, and we denote it by M. The keys of the item Z_i will be denoted by $K_{i,1}, ..., K_{i,M}$. We want to sort the items by lexicographic order on keys, i.e. Z_i will precede Z_j if there exists a number r such that

$$K_{i,s} = K_{j,s} \quad \text{for} \quad s = 1, ..., r - 1 \quad \text{and} \quad K_{i,r} < K_{j,r}.$$

We first illustrate the importance of this problem by an example. Assume that each of the items under processing has a key given by an eight-digit decadic number. We cannot use algorithm 5.4.1 because we cannot construct $100\,000\,000$ shelves. We therefore divide the key into two groups of four decadic numbers each. The lexicographic ordering according to the two resulting keys is the same as the ordering according to the values of the original key, but it requires only $10\,000$ shelves, which can be technically realised; or, we can use four keys per two decadic numbers, which requires only 100 shelves. The requirements on memory decrease significantly. The time needed for computation increases, but we shall see that the increase is not too quick.

Sorting according to multiple keys can be executed by distributing the items into shelves according to the values of the first key, then sorting each shelf according to the values of the second key, etc. This is simply a repetition of algorithm 5.4.1, but the recursion hidden in the description of the resulting algorithm can lead to difficulties of programming. There is another, less obvious but simpler, algorithm making use of the keys in reverse order. We first sort the sequence according to the values of the Mth key which is the least important for sorting. Then we sort the resulting sequence again by means of a *stable* algorithm according to the values of the $(M - 1)$th key. The resulting sorting is determined by the keys $K_{i,M-1}$, and when key values are equal the items are ordered in the same way as after the first stage, i.e. according to the keys $K_{i,M}$. In the following stages we use successively keys $K_{i,M-2}, ..., K_{i,2}, K_{i,1}$, and we finally obtain a lexicographic ordering of the sequence.

Algorithm 5.4.4. Multiple-passage sorting by distribution
Input data: numbers L and U and items $Z_1, ..., Z_n$ each of which has M integer keys $K_{i,1}, ..., K_{i,M}$ ranging from L to U inclusive.
Task: sort the items according to the lexicographic order of their keys.
Auxiliary variables: output sequence $W_1, ..., W_n$ and numbers $C_L, ..., C_U$.
   1. [*Cycle of the algorithm*] For $p := M, M - 1, ..., 1$ we successively execute step 1a.
   1a. [*pth stage*] If $M - p$ is even (i.e. if $p = M, M - 2, ...$), sort the

items $Z_1, ..., Z_n$ into the sequence $W_1, ..., W_n$ sorting them by the stable algorithm 5.4.1 according to keys $K_{i,p}$. If $M - p$ is odd, we proceed analogously, except that the items $W_1, ..., W_n$ are stored in the sequence $Z_1, ..., Z_n$.

2. If M is odd, the resulting sequence is stored in $W_1, ..., W_n$, and when needed it is transferred back to $Z_1, ..., Z_n$. ●

The speed of algorithms 5.4.1 and 5.4.4 will be discussed in the following theorem.

Theorem 5.4.5. *Algorithm* 5.4.1 *sorts* n *items with keys ranging from* L *to* U *in time* $O(n + (U - L))$. *Algorithm* 5.4.4 *sorts* n *items with* M *keys ranging from* L *to* U *in time* $O(M(n + (U - L)))$.

Proof. Algorithm 5.4.4 actually consists of M repetitions of algorithm 5.4.1. Each of the four steps of algorithm 5.4.1 is executed once, and it requires the execution of either n or $U - L + 1$ blocks of instructions each of which requires a constant time. ●

Finally, we are going to present a generalisation of algorithm 5.4.4 which will be needed later in Chapter 6. We shall now assume that the number of keys of the individual items is variable. The number of keys of the item Z_i will be denoted by M_i, and the keys themselves by $K_{i,j}$ for $j = 1, ..., M_i$. The items will again be sorted into lexicographic order which, in this case, means that Z_i precedes Z_j if there exists a number r such that $M_i \geq r - 1$, $M_j \geq r$, $K_{i,s} = K_{j,s}$ for $s = 1, ..., r - 1$ and either $M_i = r - 1$, or $K_{i,r} < K_{j,r}$.

If all the key values range from L to U, we can supply new keys $K_{i,j}$ for $j = M_{i+1}, ..., M$ where $M = \max(M_1, ..., M_n)$, by using the value $L - 1$, and then apply algorithm 5.4.4. The computation can be speeded up considerably if we compare only the actual key values, and not the fictitious value $L - 1$.

We are going to use two tricks for speeding up the computation. The first one is sorting the items according to the number of keys, i.e. in such a way that the sequence M_i becomes non-increasing. Thus, the items with a large number of keys will be found at the beginning of the sequence. Then we shall use a procedure resembling algorithm 5.4.4 which, however, will, at the jth stage, process only the items which have at least j keys. We take advantage of the fact that these items are concentrated at the beginning of the sequence.

The other improvement is based on the observation that in situations in which the algorithm will be applied, the values of the jth keys, for each individual j, are going to be rather sparse in the interval from L to U. We shall achieve an acceleration of the computation by first finding out which shelves are going to be occupied at the jth stage, and then considering only the occupied shelves in the final step of the jth stage when shelves are transferred to the output sequence.

This idea will be realised by first listing all pairs (j, k) such that there exists an item with the jth key of value k, and then sorting these pairs by a two-passage application of algorithm 5.4.4. With this information we can easily generate, for each j, an increasing sequence of numbers which are the values of the jth keys.

A realisation of these improvements can be more easily described using linked lists.

Algorithm 5.4.6. Lexicographic sorting of items with varying number of keys by distribution

Input data: numbers L and U and items $Z_1, ..., Z_n$ with Z_i having keys $K_{i,1}, ..., K_{i,M_i}$ ranging from L to U inclusive.

Task: sort the items according to the lexicographic order of their keys.

Auxiliary variables: the following linked lists:

SEQ   a working list,

KEYS(j)   for $j = 1, ..., M = \max(M_1 ... M_n)$, the elements of which are lists of values of the jth keys,

SHELF(k) for $k = L, ..., U$, the kth shelf,

CONT(j)   for $j = 1, ..., M$, containing the numbers which are values of the jth keys,

SEQ1   another working list,

SHELF1(j) for $j = 1, ..., M$, and SHELF2(k) for $k = L, ..., U$, the shelves used for creating CONT(j).

The elements of the lists SEQ, KEYS(j), and SHELF(k) are items of the same type as Z_i, the elements of SEQ1 and SHELF2(k) are pairs of integers, and SHELF1(j) and CONT(j) contain integers.

1. [*Initiation*] Put $M := \max(M_1, ..., M_n)$ and $L := M_1 + ... + M_n$. All the auxiliary lists are empty.

2. [*Specification of the values of keys*] For $i := 1, ..., n$ and $j := 1, ..., M_i$ add the pair $(j, K_{i,j})$ to the list SEQ1.

3. [*Distribution of SEQ1 according to the second key*] While SEQ1 is non-empty, execute the following instruction: delete a pair (j, k) from SEQ1, and add it to the list SHELF2(k).

4. [*Concatenation of the contents of SHELF2*] (At this moment, the list SEQ1 is empty.) For $k := L, ..., U$ concatenate the list SHELF2(k) to the end of SEQ1.

5. [*Distribution of SEQ1 according to the first key*] While *SEQ1* is non-empty, execute the following instruction: delete a pair (j, k) from *SEQ1*, and add the number k to the end of the list *SHELF*1(j).

6. [*Transfer of SHELF1(j) into CONT(j) excluding repetition*] For $j := 1, ..., M$ execute step 6a.

6a. Put $r := D - 1$ and while *SHELF*1(j) is non-empty, execute the following two instructions:

delete a number from the beginning of *SHELF*1(j), denote it by k;

if $k \neq r$, store k at the end of the list *CONT*(j) and put $r := k$.

7. [*Distribution of items according to keys*] For $i := 1, ..., n$ insert Z_i into the list *KEYS*(M_i).

8. [*Cycle of the actual sorting*] For $j := M, M - 1, ..., 2, 1$ execute steps 8a, b and c.

8a. [*Concatenation of items with j keys*] The list *KEYS*(j) is concatenated to the beginning of the list *SEQ*. (This can be performed by concatenating *SEQ* at the end of *KEYS*(j) and then transferring the new contents of *KEYS*(j) back to *SEQ*.)

8b. [*Distributing SEQ according to the jth key*] While *SEQ* is non-empty, execute the following operation: delete the first item from *SEQ*, and if the value of its jth key is k, add it at the end of *SHELF*(k).

8c. [*Collecting the shelves into SEQ*] While *CONT*(j) is non-empty, execute the following:

delete a number from the beginning of *CONT*(j), denote it by k;

concatenate the contents of *SHELF*(k) at the end of *SEQ* and then empty out *SHELF*(k) (if this is necessary in the given implementation).

9. [*Termination*] *SEQ* contains the sorted sequence of input items. ●

Theorem 5.4.7. *Algorithm 5.4.6 sorts an input sequence in time* $O(B + C)$ *where* $B = M_1 + ... + M_n$ *is the sum of the numbers of keys of the items to be sorted and* $C = U - L + 1$ *is the number of the possible values of keys.*

Proof. Both the numbers n, the number of items, and M, the maximum number of keys per item, are smaller or equal to B. Step 1 requires time $O(n + M + C)$ and steps 2 and 3 take time $O(B)$ each, since *SEQ1* contains at most B elements. Step 4 requires time $O(C)$ and steps 5 and 6, with all the executions of 6a, require time $O(B)$. Step 7 takes time $O(n)$. Step 8a is executed M times, and each execution requires constant time. The length of the sequence *SEQ* in step 8b processing the jth key is equal to the number of items with at least j keys. Denote this number by B_j. Since

$B_1 + B_2 \ldots + B_M = B$, it is clear that all the executions of step 8b together require time $O(B)$. This time will also be sufficient for step 8c, since the sum of lengths of the sequences $CONT(j)$ does not exceed B. ●

Exercises

1. Determine which of the algorithms in this chapter are stable.
2. Determine which of the algorithms in this chapter is the best one from the point of view of required memory.
3. Write down a detailed program for the algorithm Heapsort.
4. Which of the permutations of the numbers $0, 1, \ldots, 9$ will be sorted by Heapsort most quickly?
5. For each natural number n determine the permutation of the numbers $1, \ldots, n$ for which algorithm 5.2.1 needs the longest time when k in step 2 is chosen as follows:
   (a) the key of Z_i for $i = (L + R)/2$,
   (b) the mean of keys of the items with indices $(3L + R)/4$, $(L + R)/2$ and $(L + 3R)/4$,
   (c) the key of the item Z_L,
   (d) the key of the item Z_R.
6. What is the computation time of algorithm 5.2.1 with k chosen according to one of the possibilities 5(a)–5(d) above in the case where the input sequence is actually sorted, in other words, it is the sequence $1, \ldots, n$?
7. Prove that, in theorem 5.3.3, the lower bound $\log n!$ holds even for the number of executions of instructions of conditional jumps.
8. Present a method for finding, in time $O(n \log n)$, all items of a given n-item sequence which appear more than once in the sequence.
9. Present an algorithm which sorts the vertices of a graph according to their degrees in time $O(n + m)$, where n is the number of vertices and m the number of edges.

6

Problems solvable in polynomial time

The aim of this chapter is to describe the solution of some important problems for which algorithms are known that process even wide ranges of input data in a sufficiently short time. The speed of those algorithms is satisfactory even in the worst cases of computation.

6.1 Paths in graphs

Throughout the present section all graphs, edges, paths, etc, are meant to be directed. We will be interested in the determination of distances between vertices of an edge-labelled graph.

Definition 6.1.1. Let G be an edge-labelled graph. By the length of a directed path in G we mean the sum of the evaluations of all edges of the path. The distance of a vertex w from a vertex v is defined to be the least length of a path leading from v to w in the graph G, if such exists; else, the distance is equal to $+\infty$. The distance of each vertex from itself is zero. ●

Let us remark that the length of a path as defined above does not correspond to the number of edges of the path. In the present section, the label of an edge will be called its *length*.

In applications of the algorithms which we are going to describe here the label of an edge or a path need not mean the actual distance, but, for example, the time needed for the transition from one vertex to another, or the cost of such a transition. Sometimes the edges will represent certain activities which are to be performed, and the label will mean the time required for the performance, and so on.

We are going to investigate the following problems.
1. Given vertices u and v, determine the distance from u to v.
2. Given a vertex v, determine the distances of all vertices of the graph G from v.

3. Given a vertex u, determine its distance from all vertices of the graph G.
4. Determine the distances of all pairs of vertices of the graph G.

Task 3 can be easily converted to task 2 by reverting the direction of all edges while keeping their lengths. We are, therefore, not going to present special algorithms for problem 3.

It turns out that all the known procedures for solving problem 1 generate, as a byproduct, distances of u from all vertices of the graph (or distances of all vertices from u), and they thus represent algorithms solving problems 2 and 3 too. The only difference is the possibility of terminating the computation of problem 1 at the moment when the distance from u to v is determined.

Below we restrict ourselves to the following two special types of labelled graphs: we assume either that the lengths of edges are all non-negative, or that the graph does not have cycles. The reasons for this restriction will be made clear in the exercises of Chapter 7.

The simplest case of problem 2 is when the length of each edge is equal to 1, since then the length of a path is just the number of edges of which it consists. In that case the solution for undirected graphs can be found by algorithm 4.2.5, and for the directed ones by algorithm 4.2.9. By theorem 4.2.2 (and the analogous result for directed graphs), the computation takes time $O(n + m)$ for a graph of n vertices and m edges.

Breadth-first searching can, however, lead to an incorrect result for a general labelled graph even if the labels are non-negative. In that case we can use the following classical procedure designed by E W Dijkstra [82], which, essentially, also searches the given graph.

Algorithm 6.1.2. Determination of distances from a given vertex of a non-negatively labelled graph

Input data: a non-negatively labelled directed graph G and its vertex u.
Task: for each vertex v of the graph G determine the distance d_v from u to v.
Auxiliary variables:

M, the set of all vertices v for which the distance d_v has not yet been definitely specified;

V_v, for each vertex v a vertex V_v is specified as the last-but-one vertex of the shortest path from u to v which does not pass through any vertex of the set M;

L_{vw}, the direct distance from v to w, for arbitrary vertices v and w.

1. [*Initiation*] If $v = w$, put $L_{vw} := 0$;
if (v, w) is an edge of G, put $L_{vw} :=$ the length of the edge (v, w);
if (v, w) is not an edge of G, put $L_{vw} := +\infty$.

$M :=$ the set of all vertices of G distinct from u; $d_u := 0$; $V_u := u$; if v is a vertex of G distinct from u, put $d_v := L_{uv}$ and if $d_v < +\infty$, put $V_v := u$.

2. [*Termination test*] If the set M is empty, the computation is terminated.

3. [*Determine d_v for the next vertex*] We choose the vertex of M having the least value of d_v (choosing arbitrarily when there are several possibilities). If $d_v = +\infty$, the computation is terminated; else, we delete v from the set M.

4. [*Actualise d_w and V_w*] If v is the vertex chosen in step 3, then for each vertex w of the set M we do the following:

if $d_v + L_{vw} < d_w$, put $V_w := v$ and $d_w := d_v + L_{vw}$.

5. Jump to step 2. ●

Dijkstra's algorithm can work more slowly than the procedure based on breadth-first searching when the number of vertices of the graph G is small.

Theorem 6.1.3. *Algorithm 6.1.2 will process a graph of n vertices in running time $O(n^2)$.*

Proof. Step 1 takes running time $O(n^2)$. Each execution of steps 2 to 5 will decrease the set M by an element, and hence, they will be executed $n - 1$ times. The determination of the minimum in step 3 and searching the vertices in step 4 can be performed in running time $O(n)$, and steps 2 and 5 require constant time. ●

It is clear from the above proof that the running time estimate of theorem 6.1.3 cannot be improved. No algorithm is presently known which would compute more quickly from the asymptotic point of view, e.g. in running time proportionate to the sum of the number of edges and vertices as in the case of breadth-first searching. However, if the set M of 6.1.2 is implemented as a priority queue (which speeds up step 3), we can get an $O(m \log n)$ time algorithm (where m is the number of edges) which is faster for sparse graphs.

The proof of correctness of algorithm 6.1.2 is somewhat more complicated.

Theorem 6.1.4. *At the moment of termination of algorithm 6.1.2, the distance from u to any vertex v is equal to d_v, and the shortest path from u to v, if $d_v < +\infty$, is the following: $u = x_0, x_1, \ldots, x_k = v$ with $x_{i-1} = V_{x_i}$ $(i = 1, \ldots, k)$.*

Proof. We will say that, at a certain stage of algorithm 6.1.2, a path

$u = y_0, y_1, \ldots, y_j$ is final if none of the vertices $y_1, y_2, \ldots, y_{j-1}$ lies in the set M. For each vertex v denote by d_v the distance from u to v in the graph G. We shall prove that in each entry into step 2 the following hold:

(a) if $z \notin M$ and $w \in M$, then $d_z \leq d_w$,
(b) if $d_w = +\infty$ then there does not exist a final path from u to v,

and

(c) if $d_w < +\infty$, then d_w is the length of the shortest final path from u to w, and we have $d_V + L_{V_w w} = d_w$.

The statement is certainly true at the moment of initiation because then the only final path into a vertex w can be the single edge (u, w). We are going to verify that the statement remains valid after each execution of steps 3 and 4.

Let v be the vertex chosen in step 3. Given vertices $z \notin M$ and $w \in M$, we have $d_z \leq d_v \leq d_w$ and, hence, the execution of step 3 does not destroy the validity of the statement (a); if in step 4 some d_w is changed, then the new value fulfils $d_v \leq d_v + L_{vw} = d_w$, due to the assumption that the evaluation is non-negative.

By deleting the vertex v from the set M, new final paths might have been created. Naturally, no such path leads into the vertex v. We will now show that no such final path leading to a vertex $z \notin M$ has length smaller than d_z: such a path would necessarily pass through the vertex v, and its length would have to be at least d_v, but $d_z \leq d_v$.

Let $u = y_0, y_1, \ldots, y_j = w$ be a new final path of length smaller than the value d_w at the instant before the considered execution of steps 3 and 4. It follows from the above that $w \neq v$ and $w \in M$. Since either $y_{j-1} = v$ or $y_{j-1} \notin M$, it follows from the preceding paragraph that the path $y_0, y_1, \ldots, y_{j-1}$ has length larger than or equal to $d_{y_{j-1}}$. Assuming $y_{j-1} \neq v$, then we could substitute the section $y_0, y_1, \ldots, y_{j-1}$ of the path by the path from u to y_{j-1} which would have had the length $d_{y_{j-1}}$ and would have been final already before the vertex v was deleted from M — but this would mean that before v was deleted from M there had existed a final path from u to w (passing through y_{j-1}) of length less than d_w, which contradicts the statement (c), valid before deleting v.

Thus, we conclude that $y_{j-1} = v$, and since the length of the path $y_0, y_1, \ldots, y_j$ is at least $d_v + L_{vw}$, it follows that $d_v + L_{vw} < d_w$. Consequently, the value of d_w is going to be changed in step 4 so as not to be larger than the length of the new final path.

Conversely, if $d_v + L_{vw} < d_w$, then certainly $L_{vw} < +\infty$, and hence (v, w) is an edge of the graph G. By concatenating the shortest final path from u to v and the edge (v, w), we obtain a path from u to w of length $d_v + L_{vw}$, which becomes final after deleting v from the set M.

We now see that the algorithm proceeds in such a way that the statements (b) and (c) above also remain valid. After termination of the computation each path is final. In fact, if there existed a non-final path $u = y_0, y_1, ..., y_j$, then we could choose the smallest member p such that $y_p \in M$, and then $y_0, y_1, ..., y_p$ would be a final path. Then (c) above implies that $d_{y_p} < +\infty$. Clearly, $p < j$, and while a vertex p with such properties exists, the algorithm cannot terminate. Thus, after termination of the computation, the statements (b) and (c) hold for all paths, and consequently, $d_v = D_v$ for all vertices v.

The above considerations make it possible to write down a recursive formula for the shortest (final) path from u to v. ●

Another well-known algorithm for finding the shortest path is due to R Bellman [56], and it uses methods of dynamic programming. A survey of the whole field of problems connected with searching the shortest path can be found in the paper [85] of S E Dreyfus, see also [108, 109, 161, 234, 236, 264].

By applying Dijkstra's algorithm we can also solve problem 4 above, the determination of the matrix of distances of a non-negatively labelled graph. We proceed by computing, for each vertex u, distances from u to all vertices of the given graph. This requires time $O(n^3)$. There is a more efficient algorithm designed by R W Floyd [105].

Algorithm 6.1.5. Computation of the matrix of distances of vertices
Input data: a non-negatively evaluated graph G with vertices $v_1, v_2, ..., v_n$.
Task: for each pair i and j determine the distance d_{ij} from the vertex v_i to the vertex v_j in the graph G.
 1. [*Initiation*] For all $i, j = 1, 2, ..., n$ initiate as follows:
if $i = j$, put $d_{ij} := 0$,
if (v_i, v_j) is an edge of the graph G, put $d_{ij} :=$ the length of the edge (v_i, v_j),
if $i \neq j$ and (v_i, v_j) is not an edge of the graph G, put $d_{ij} := +\infty$.
 2. [*Cycle*] For $k := 1, 2, ..., n$ execute 2a.
 2a. For each pair $i, j = 1, 2, ..., n$ with $i \neq j$ put $d_{ij} := \min(d_{ij}, d_{ik} + d_{kj})$.
●

Theorem 6.1.6. *Algorithm 6.1.5 will process a graph G of n vertices in time $O(n^3)$. After termination, each of the numbers d_{ij} is equal to the distance from the vertex v_i to the vertex v_j in the graph G.*

Proof. The time estimate certainly needs no explanation. A proof of the proper function of the algorithm follows from the fact that after each of the executions of step 2a there is a number k such that d_{ij} is equal to the length of the shortest path from v_i to v_j which, besides the two vertices, contains only vertices of the set $v_1, v_2, ..., v_k$. ●

Although it may seem that the estimate $O(n^3)$ cannot be improved, this is not true. The computation of the matrix of distances is strongly related to the algorithms for a quick multiplication of matrices (see [103, 211] and the exercises below). In that field surprising results have been obtained. For example, an algorithm multiplying integer matrices of order n in time $O(n^{2.49})$ has been found, and a simple method requiring time $O(n^{2.79})$ has been presented, see [1, 247]. On the basis of those algorithms, a method for computing the matrix of distances in time $O((n \log \log n/\log n)^3) = o(n^3)$ has been given; see [78]. The speeding up of the algorithm is, however, so small that its effect is observable only for large values of n, and the algorithm is substantially more complicated than algorithm 6.1.5 which, therefore, seems more suitable in practice.

We now turn to another important case in which the shortest path connecting two given vertices can be found easily: we shall investigate edge-labelled acyclic directed graphs. It can be shown that Dijkstra's algorithm cannot be applied to that case because it can yield wrong results. On the other hand, the algorithm we present below makes use of the absence of cycles to perform computation more quickly than Dijkstra's algorithm.

Let us remark that by changing the sign of the evaluation of edges, the shortest path becomes the longest one, and vice versa. Since we no longer require that labelling be non-negative, the algorithm below can thus be used also for the computation of the longest path between two given vertices which is the foundation of the well-known methods CPM and PERT.

We proceed by first ordering the vertices of the graph into a sequence $v_1, v_2, ..., v_n$ such that each directed edge (v_i, v_j) fulfils $i < j$. We shall show that this ordering of vertices, called topological, under which all edges are directed from left to right, exists if, and only if, the graph is acyclic, and we shall describe a way of finding the ordering.

A topological ordering can be found as follows: choose as v_1 a vertex having no incoming directed edges. Having chosen vertices $v_1, ..., v_{i-1}$, the vertex v_i is chosen in such a way that it has no incoming edges in the full subgraph of the given graph formed by deleting the edges $v_1, ..., v_{i-1}$. Since the importance of this procedure goes beyond the problem of path searching, we describe it as a separate algorithm.

Algorithm 6.1.7. Topological ordering of a graph

Input data: a directed graph G of n vertices.

Task: sort the vertices of the graph G into a sequence $v_1, v_2, ..., v_n$ such that each directed edge (v_i, v_j) of the graph G fulfils $i < j$.

Auxiliary variables: for each vertex v a number C_v is given, expressing the number of vertices not yet searched, from which an edge leads into v. A set M is used containing the vertices v not yet searched for which $C_v = 0$.

1. [*Determine C_v*] First, for each vertex v, put $C_v := 0$, and then for each directed edge (v, w) put $C_w := C_v + 1$.

2. [*Initiate M*] Insert all vertices v with $C_v = 0$ into M.

3. [*Initiate the ordering index*] $i := 0$.

4. [*Test of cycles*] If $M = \emptyset$ and $i < n$, the computation terminates: the graph G contains a cycle, and a topological ordering of vertices does not exist.

5. [*Determine the next vertex*] Choose an arbitrary vertex v in M. Delete v from the set M, put $i := i + 1$ and denote v by v_i.

6. [*Termination test*] If $i = n$, the computation is terminated, the ordering of vertices has been found.

7. [*Actualise C_w and M*] If v is the vertex deleted in step 5, then for each outcoming directed edge (v, w) put $C_w := C_w - 1$ and if C_w becomes zero, insert the vertex w into the set M.

8. Jump to step 4.

Theorem 6.1.8. *Algorithm 6.1.7 will process a graph of n vertices and m edges in running time $O(n + m)$, and it will find a topological ordering of vertices if, and only if, the graph does not contain a cycle.*

Corollary 6.1.9. *The vertices of a directed graph can be topologically ordered if, and only if, the graph is acyclic.*

Proof. For each vertex v, the following operations are performed at most once: putting $C_v := 0$ in step 1, testing whether C_v is zero and, eventually, inserting v into the set M in step 2, deleting v from M in step 5, and inserting v into M in step 7. For each edge (v, w), the following operations are performed at most once: $C_w := C_w + 1$ in step 1, and $C_w := C_w - 1$ and the test whether C_w is zero in step 7. Steps 1, 2 and 3 are performed once, steps 4 to 8 at most $n + 1$ times, and hence, the remaining procedures of the algorithm require running time $O(n)$.

It is clear that the variables C_v and M are created and kept in such a way that C_v is equal to the number of edges leading into the vertex v from the vertices the order of which has not yet been specified. Further, M contains the vertices not yet ordered for which $C_v = 0$. If (v, w) is an edge of the graph, then w cannot be inserted into M before the vertex v has been denoted, and hence, v must precede the vertex w.

Let G be acyclic. We want to prove that until the order of all vertices is specified, the set M is non-empty: if this statement were false, then at some stage of the computation each vertex not yet denoted would have an incoming edge whose initial vertex has also not yet been denoted. Then we would obtain an infinite sequence $z_1, z_2, z_3, \ldots$ of vertices not yet denoted such that the pairs $(z_2, z_1), (z_3, z_2), \ldots$ are edges of the graph G. Since the set of vertices of the graph G is finite, such a sequence necessarily contains a cycle, which is a contradiction.

Conversely, if G contains a cycle, then a topological ordering of vertices clearly does not exist. ●

We now describe an algorithm for searching paths in an acyclic graph.

Algorithm 6.1.10. Finding extremal paths in an acyclic graph
Input data: an edge-labelled graph G with a vertex u.
Task: verify that G is acyclic, and if so, determine the lengths of the shortest or the longest paths from u to all vertices of G.
Auxiliary variables: a number d_v for each vertex v.

1. [*Topological ordering of vertices*] Using algorithm 6.1.7, find a topological ordering $v_1, v_2, \ldots, v_n$ of the vertices, or prove that such an ordering does not exist. In the latter case terminate the computation.

2. [*Initiation*] Put $d_u := 0$,
if (u, v_i) is an edge of the graph G, put $d_{v_i} :=$ the length of the edge (u, v_i),
if (u, v_i) is not an edge of G, and $u \neq v_i$, put

$$d_{v_i} = \begin{cases} +\infty & \text{if the shortest path is to be found} \\ -\infty & \text{if the longest path is to be found.} \end{cases}$$

3. [*Determine d_v*] Assuming that $u = v_p$, then for $i := p, p+1, \ldots, n$ execute the following operations for each outcoming edge (v_i, v_j) of the vertex v_i:
$d :=$ the length of the edge (v_i, v_j),

$$d_{v_j} := \begin{cases} \min(d_{v_j}, d_{v_i} + d) & \text{if the shortest path is to be found} \\ \max(d_{v_j}, d_{v_i} + d) & \text{if the longest path is to be found.} \end{cases} \quad ●$$

Theorem 6.1.11. *Algorithm 6.1.10 will process a graph of n vertices and m edges in running time $O(n + m)$. After termination, the number d_v of any vertex v is equal either to $\pm\infty$, when there does not exist a path from u to v, or to the shortest or the longest path from u to v, if such a path exists.*

Proof. Step 1 takes running time $O(n + m)$, see 6.1.8, and step 2 takes running time $O(n)$. In step 3 each vertex and each edge is processed at most once, and hence running time $O(n + m)$ is sufficient.

Analogously to the above proofs of correctness of Dijkstra's and Floyd's algorithms, it is possible to show that at the instant of terminating step 3 for processing a certain i, the number d_v is equal, for all $u = u_p, u_{p+1}, ..., u_n$, to the shortest length of a path from u to v which contains, besides those two vertices, only vertices from the set $\{u_{p+1}, ..., v_i\}$ (or equal to $\pm\infty$, if such a path does not exist), and for the vertices u_j, where $j < p$, the numbers d_{v_j} are $\pm\infty$. ●

6.2 Minimum spanning tree of a graph

In the present section we investigate positively edge-labelled, connected, undirected graphs. To stress the motivation, the labelling of the edge will be called *price*. Our goal is to find a collection of edges which would connect all vertices as cheaply as possible. In other words, we are looking for a subgraph H of the given graph G which has the same set of vertices as G, is connected, and has the least possible sum of prices of its vertices.

The type of such a graph is described by the following lemma.

Lemma 6.2.1. *Let G be a positively edge-labelled connected, undirected graph. Then there exists a graph H with the following properties:*

(a) *H is a connected subgraph of G with the same set of vertices as G,*

(b) *if K is a connected subgraph of G with the same set of vertices as G, then the sum of labels of all edges of the graph H is smaller than or equal to the sum of labels of all edges of K. Moreover, each graph H with the above properties (a) and (b) is a tree.*

Proof. The set M of all connected subgraphs of the graph G is finite and non-empty, since it contains G itself. Therefore, there is a graph in M with the least sum of evaluation of edges. If H is such a graph, then H cannot contain a cycle since by deleting an edge from a cycle of H we would obtain a graph obviously lying in the set M, which, however, has a smaller sum of labels than the graph H — a contradiction. Thus, H is a tree. ●

A graph H satisfying conditions (a) and (b) of the above lemma is called a *minimum spanning tree* of the graph G. An algorithm for searching a minimum spanning tree was described as early as 1926 by O. Borůvka [65] and later by G Choquet [71], but it was not until J B Kruskal published his paper [183] in 1956 (i.e. in the period in which computers were already available) that the procedure became generally known. The whole algorithm is quite simple. The edges of the graph *G* are first ordered in such a way that their prices are non-decreasing, then they are searched in that ordering, and a vertex is inserted in the graph *H* if, and only if, it does not create a cycle with the edges already in *H*. As an example, consider figure 18 in which the minimum spanning tree is denoted by bold lines. The edges are searched in the following order:

$$\underline{ST}, \ \underline{UZ}, \ \underline{UV}, \ VZ, \ \underline{SX}, \ \underline{WZ}, \ VW, \ \underline{TZ}, \ TU, \ WX, \ \underline{WY}, \ TY, \ XY,$$

and the underlined ones are inserted into *H*, whereas the remaining edges are not inserted into *H* because they form a cycle with some of the left-hand underlined edges.

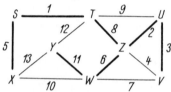

Figure 18.

An edge *h* forms a cycle with the edges already inserted in *H* if, and only if, its end-vertices lie in the same component of the current form of the graph *H*. (For example, in figure 18 at the instant of searching the edge *TZ*, we have the following components: $\{S, T, X\}$, $\{U, V, W, Z\}$ and $\{Y\}$, and we therefore insert the edge *TZ*, thus creating a new component $\{S, T, U, V, W, X, Z\}$. Consequently, the edge *WX* is not inserted into *H*.) This is used in the description of the Borůvka–Choquet–Kruskal algorithm in which the decomposition of the graph *H* into components is stored in the form of a factor set.

Algorithm 6.2.2. Determination of a minimum spanning tree of a graph
Input data: a positively edge-labelled, connected graph *G*.
Task: find a minimum spanning tree *H* of the graph *G*.
Auxiliary data structure: a factor set on the vertices of the graph *G*.
  1. [*Ordering of edges*] The edges of *G* are ordered according to their labels in such a way that the labels are non-decreasing.
  2. [*Initiation*] Put $V(H) := V(G)$ and $E(H) = \emptyset$. The factor set expresses the decomposition of $V(H)$ into singleton sets.
  3. [*Insert an edge*] The edges of the graph *G* are searched in the order given in step 1, and for each edge $h = \{v, w\}$ the following operations are performed:
  by executing $FIND(v)$ and $FIND(w)$ in the factor set we determine

the components V and W of the graph H in which the vertices v and w, respectively, are lying; if $V \neq W$, then h is inserted in $E(H)$ and $UNION(V, W)$ is executed. ●

We postpone the proof of correctness of the above algorithm, and just mention its speed.

Theorem 6.2.3. *Algorithm 6.2.2 will process a graph of n vertices and m edges in running time $O(m \log n)$.*

Proof. A suitable implementation of step 1 requires running time $O(m \log m) = O(m \log n)$, since $\log m \leq \log n^2 = 2 \log n$. The $2m$ executions of the operation $FIND$ and the $n - 1$ executions of the operation $UNION$ require substantially less time, as shown in § 3.10. The remaining activities of the algorithm require running time $O(m)$ because a connected graph has the number of vertices larger at most by 1 than the number of edges. ●

The most time-consuming operation of the above algorithm is the ordering of edges. The operation $FIND$ of the factor set can, as shown in exercise 12 of Chapter 3, be performed without path shortening and this would not change the estimate of theorem 6.2.3. However, if the evaluation of edges makes it possible to use sorting by distribution and to perform step 1 in time $O(m)$, then we obtain an algorithm the speed of which is determined by the speed of execution of the operations on factor sets, and this speed cannot be obtained by any other procedure known today.

For graphs with a large number of edges (of order n^2) there is another procedure which may be faster and which, by design and speed, resembles Dijkstra's algorithm of the preceding section. That procedure was described by V Jarník [160] as early as 1930, but it did not become well known until the publication of R C Prim's paper [221].

Algorithm 6.2.4. Determination of the minimum spanning tree of a graph
Input data: positively edge-labelled connected graph G.
Task: find a minimum spanning tree H of the graph G. Auxiliary variables:
    M, the set of vertices not yet covered by the graph H,
    variable V_v, for each vertex v, which is either undefined, or equal to the vertex to which the vertex v was connected in the course of generation of the graph,
    number d_v, equal to the length of the vertex connecting v with V_v, in the case when V_v is defined.
    1. [*Initiation*] Choose an arbitrary vertex v of the graph G. Set

$V(H) := \{u\}$, $E(H) := \emptyset$, $M := V(G) - \{u\}$, and for each vertex $v \in M$ do the following:

if (u, v) is a vertex of the graph G, put $V_v := u$ and $d_v :=$ the evaluation of $\{u, v\}$;

else, V_v is undefined and $d_v := +\infty$.

2. [*Termination test*] If $M = \emptyset$, the computation is terminated.

3. [*Determine the next edge of* H] Choose the element v of the set M for which d_v is minimum. Insert v into $V(H)$ and $\{v, V_v\}$ into $E(H)$, and delete v from the set M.

4. [*Actualise* d_w *and* V_w] For each vertex v of the set M do the following: if (v, w) is an edge of the graph G with evaluation smaller than d_w, then set $V_w := v$ and $d_w :=$ the evaluation of $\{v, w\}$.

5. Jump to step 2. ●

The proof of correctness of the above algorithm will also be postponed. This algorithm strongly resembles that of Dijkstra, except that the number d_v does not represent the distance from the vertex v to u, but only to the nearest vertex of the graph H, and this influences the role of V_v. We therefore leave the proof of the following theorem to the reader as an exercise.

Theorem 6.2.5. *Algorithm 6.2.4 will process a graph of n vertices in running time* $O(n^2)$. ●

Both of the above procedures can be understood as special cases of the following general one.

Algorithm 6.2.6. General scheme for finding the minimum spanning tree of a graph

Input data: a positively edge-labelled connected graph G.

Task: find a minimum spanning tree H of the graph G.

Auxiliary variable: a subgraph L of the graph G not containing cycles (a forest).

1. [*Initiation*] $V(L) := V(G)$ and $E(L) := \emptyset$.

2. [*Termination test*] If L is a tree, set $H := L$ and terminate the computation.

3. [*Choose a component*] Choose an arbitrary component C of the graph L.

4. [*Determine the nearest component*] Choose an edge $\{v, w\}$ with $v \in C$ and $w \notin C$ with the minimum label. Denote by D the component containing the vertex w.

5. [*Connect components*] Insert the edge chosen in step 4 to $E(L)$. (This connects the components C and D of the graph L into one.)
   6. Jump to step 2.   ●

In the procedure of algorithm 6.2.2 the graph L is identical with the graph H, and the determination of the component C consists of first choosing the edge $\{v, w\}$ and then denoting by C the component in which the vertex v lies. In order to guarantee that the edge $\{v, w\}$ has the required properties of algorithm 6.2.6, it is necessary to order the edges in the first place.

In algorithm 6.2.4, one of the components of L is the generated graph H, and the other components are singleton sets. Algorithms 6.2.4 and 6.2.6 are closely connected.

We now verify the correctness of the general scheme, and hence, of the preceding two algorithms too.

Theorem 6.2.7. *Algorithm 6.2.6 terminates after a finite number of steps, and after termination, the graph H is a minimum spanning tree of the graph G.*

Proof. The graph L does not contain cycles and hence were the algorithm never to stop, L would always have at least two components, which would imply that there is an edge in the graph G which has end-vertices in different components of the graph L. The number of such edges decreases by each execution of steps 4 and 5, and hence the cycle of steps 2 to 5 of the algorithm is executed at most $n - 1$ times, where n is the number of vertices of the graph G.

We now prove that during the entire computation the following statement remains valid: there exists a minimum spanning tree of the graph G containing the graph L as a subgraph.

This statement is certainly true at the moment of initialisation because then $E(L) = \emptyset$. Next, assume that before step 5 is executed, the graph L is a subgraph of a minimum spanning tree K of the graph G. If the inserted edge $h = \{v, w\}$ is also an edge of the spanning tree K, then the execution of step 5 does not ruin the validity of the above statement. Assume that h is not an edge of the spanning tree K. Assume, further, that $v \in C$ and $w \notin C$, where C is the component of the graph L chosen in step 3. On the uniquely determined path from v to w in the graph G there must lie an edge $k = \{x, y\}$ such that $x \in C$ and $y \notin C$, and thus an edge which is not an edge of the graph L. It follows from the choice of the edge h that the label of k is larger than or equal to that of h. The edge h together with the path from v to w form a cycle. Therefore, by deleting the edge k from the graph K and inserting the edge h we obtain a graph K_1 which is again connected, and contains all vertices of the graph G. Moreover, the sum of labels of the edges of K_1 is not larger than the sum of labels of the edges of K, and hence K_1 is also a minimum spanning tree of the graph G. This minimum spanning tree contains the graph L after step 5.

The above statement holds, naturally, also after termination of the computation. At that moment, L is itself a tree, and hence it has the same number of edges as the minimum spanning tree K. Consequently, L is equal to K, and since we finally set $H := L$, we conclude that, after termination, H is a minimum spanning tree of the graph G too. ●

It has turned out that the general scheme above can be used to find faster algorithms for finding a minimum spanning tree than algorithms 6.2.2 and 6.2.4 (with the exception of the case when algorithm 6.2.2 allows sorting by distribution). We are thus going to investigate the possibilities of implementation of the scheme 6.2.6 in more detail.

The decomposition of the set of vertices into components of L can be suitably stored in the form of factor set in a manner already indicated in the course of description of algorithm 6.2.2. (In algorithm 6.2.4, the special form of L even made it possible to simplify the description via the set M, but this case is not typical for the variants of algorithm 6.2.6 mentioned below.)

To make the choice of the edge $\{v, w\}$ in step 4 of algorithm 6.2.6 easier, it is suitable to create, for each component X of the graph L, a mergeable heap $Q(X)$ in which all edges with at least one end-vertex lying in X are stored. The place of a vertex in the heap is controlled by its labelling. Step 4 is then executed by choosing vertices from the heap $Q(C)$ using the operation *EXTRACT-MIN* until we find an edge with one end-vertex lying outside C.

To determine the components in which the end-vertices of the edges currently chosen are lying, we apply the operation *FIND* of factor set. In step 5 it is necessary to perform the union of the unifiable heaps $Q(C)$ and $Q(D)$.

The role of a suitable choice of the component C in step 3 is also important. Very good results can be obtained by using one of the following possibilities:

(a) the component with the minimum number of vertices is always chosen as C,

(b) a list of components of the graph L is created, and the first component on the list is always chosen as C; in step 5 both components C and D are deleted from the list and the component obtained by executing the union of C and D is inserted at the end of the list.

The advantage of procedure (a), which is absolutely opposite to the (relatively slow) algorithm 6.2.4 above, is that the operation on the heap $Q(C)$ determining the edge (v, w) will be fast, due to the small size of the heap. Procedure (b) tries not to make a preference of any of the components, and, hence, to get a uniform growth of the components, thus obtaining the same effect as (a). The advantage of this method is the simplicity of implementation.

In some cases it is advantageous to leave only those edges in G which either lie in the generated graph L, or are the shortest edges connecting components of the graph L. However, it is difficult to determine conditions under which the speed-up of the computation will prevail over the loss of time caused by the choice of all edges which are to be deleted as unsuitable.

These methods of implementation can be found in the papers of A C Yao [272] and of D Cheriton and R E Tarjan [70], where algorithms have been described which will process a graph of n vertices and m edges in running time $O(m \log \log n)$. Moreover, dense graphs, for which $m = \Omega(n^{1+\varepsilon})$, $\varepsilon > 0$, will be processed in running time $O(m)$, and planar graphs even in running time $O(n)$. Sometimes the minimal spanning tree is given, but it is necessary to prove its minimality. J Komlos [180] showed how to do this in $O(n)$ time.

6.3 Flows in networks

This section deals with a problem which has a number of applications in optimisation of transport or transmission, and which is very interesting and well developed theoretically. The whole field of problems is investigated in the books [15, 16, 26 and 24], the first of which has had a particular impact on the development of the methods.

Definition 6.3.1. A *network* is a quadruple $S = (G, s, t, c)$, where G is a directed graph, s and t are two distinct vertices of the graph G, called the *source* and *target* respectively, and c is a function assigning to each directed edge h of the graph G a positive real number $c(h)$, called the *capacity* of the edge h. ●

In the rest of §6.3 we will say 'edge' in place of 'directed edge', for simplicity. We further introduce the following convention.

Convention 6.3.2. If G is a directed graph and f a function assigning numbers to the edges of G, then for each pair v and w of vertices G we write

$$f(v, w) = \begin{cases} f(h) & \text{if } h = (v, w) \text{ is an edge of } G, \\ 0 & \text{if } (v, w) \text{ is not an edge of } G. \end{cases}$$

Definition 6.3.3. A *flow* in the network $S = (G, s, t, c)$ is a function f assigning to each edge of the graph G a number $f(h)$, called the *flow through the edge h*, in such a way that the following conditions hold.

1. For each edge h of the graph G, $0 \leq f(h) \leq c(h)$.

2. For each vertex v of the graph G distinct from the source and the target we have

$$\sum_{(u,v) \in E(G)} f(u, v) = \sum_{(v,u) \in E(G)} f(v, u).$$

The magnitude of the flow f is defined to be the number

$$|f| = \sum_{(s,v) \in E(G)} f(s, v) - \sum_{(v,s) \in E(G)} f(v, s).$$

A flow f is said to be *saturated* if for each directed path $s = v_0, v_1, ..., v_k = t$ in the graph G there exists an edge (v_{i-1}, v_i) such that

$$f(v_{i-1}, v_i) = c(v_{i-1}, v_i).$$

A *maximum flow* is a flow f such that for each flow f_1 in the given network we have $|f_1| \leq |f|$. ●

The flow in a network is thus a mathematical concept expressing movement of some medium which neither gets lost nor springs up in the inner vertices of the network, and the amount of which flowing through an edge does not exceed the capability of the edge to hold it, determined by the capacity. The magnitude of the flow expresses the amount of medium which springs up in the source, and which, as follows from condition 2 of definition 6.3.3, is then swallowed by the target. It is our aim to find a maximum flow in a network. The following classical procedure has been designed by L R Ford and D R Fulkerson [15].

Algorithm 6.3.4. Determination of a maximum flow in a network
Input data: a network $S = (G, s, t, c)$.
Task: find a maximum flow f in the network S.

1. [*Initiation*] For all edges h of the graph G put $f(h) := 0$.
2. [*Find an augmenting path*] We are looking for a sequence of vertices $s = v_0, v_1, ..., v_k = t$ such that for each $i = 1, ..., k$ the following holds: either (v_{i-1}, v_i) is an edge of G and

$$c(v_{i-1}, v_i) - f(v_{i-1}, v_i) > 0$$

(i.e. the flow through the edge can be augmented), or (v_i, v_{i-1}) is an edge of G and $f(v_i, v_{i-1}) > 0$ (i.e. the flow in the backwards direction can be decreased).

If such a path does not exist, the computation is terminated, and f is a maximum flow.

3. [*Determine the augmentation*] For $i = 1, ..., k$ put

$$d_i := (c(v_{i-1}, v_i) - f(v_{i-1}, v_i)) + f(v_i, v_{i-1}).$$

Further, put $d := \min (d_1, d_2, ..., d_k)$.
4. [*Augment the flow*] For $i = 1, ..., k$ put

$$d'_i := \min (d, c(v_{i-1}, v_i) - f(v_{i-1}, v_i))$$

and

$$d''_i := d - d'_i.$$

Put $f(v_{i-1}, v_i) := f(v_{i-1}, v_i) + d'_i$ if (v_{i-1}, v_i) is an edge of G, and $f(v_i, v_{i-1}) := f(v_i, v_{i-1}) - d''_i$ if (v_i, v_{i-1}) is an edge of G.
5. Jump to step 2. ●

Before proving the correctness of algorithm 6.3.4, we introduce an important concept.

Definition 6.3.5. By a cut in a network $S = (G, s, t, c)$ is understood a set A of vertices of the graph G which contains the source and does not contain the target. The magnitude of the cut A is the number

$$|A| = \sum_{\substack{(v,w) \in E(G) \\ v \in A, w \notin A}} c(v, w).$$

The following theorem expresses the relation between flows and cuts.

Theorem 6.3.6. *For each flow f and each cut A in a network S, we have* $|f| \leq |A|$.

Proof. We compute as follows:

$$|A| = \sum_{\substack{(v,w)\in E(G)\\ v\in A,\, w\notin A}} c(v,w) \geq \sum_{\substack{(v,w)\in E(G)\\ v\in A,\, w\notin A}} f(v,w) - \sum_{\substack{(w,v)\in E(G)\\ v\in A,\, w\notin A}} f(w,v)$$

$$= \sum_{v\in A}\Big(\sum_{(u,v)\in E(G)} f(u,v) - \sum_{(v,w)\in E(G)} f(v,w)\Big)$$

$$= \sum_{(u,s)\in E(G)} f(u,s) - \sum_{(s,w)\in E(G)} f(s,w) = |f|. \quad \bullet$$

We are now prepared to prove the correctness of the Ford–Fulkerson algorithm.

Theorem 6.3.7. *After termination of the computation of algorithm 6.3.4, the flow f is a maximum flow of the network S.*

Proof. It is not difficult to verify that the choice of the number d and the modifications of the function f in step 4 of algorithm 6.3.4 are performed in such a way that the function f remains a flow throughout the computation.

Let f_0 denote the value of f after the termination of the computation. To verify that the magnitude of f_0 is maximum, we denote by A the set consisting of the source and of all vertices v such that there exists a sequence of vertices $s = v_0, v_1, ..., v_k = v$ with the following property:

$$c(v_{i-1}, v_i) - f_0(v_{i-1}, v_i) + f_0(v_i - v_{i-1}) > 0 \quad \text{for all } i = 1, 2, ..., k.$$

It follows from the condition of termination of the algorithm that A does not contain the target and hence A is a cut. For each edge $(v, w) \in E(G)$ with $v \in A$ and $w \notin A$ we necessarily have $c(v, w) - f_0(v, w) + f_0(w, v) = 0$, and it follows that

$$c(v, w) = f_0(v, w) \quad \text{and} \quad f_0(w, v) = 0.$$

Consulting the proof of theorem 6.3.6, we conclude that

$$|A| = \sum_{\substack{(v,w)\in E(G)\\ v\in A,\, w\notin A}} c(v,w)$$

$$= \sum_{\substack{(v,w)\in E(G)\\ v\in A,\, w\notin A}} f_0(v,w) - \sum_{\substack{(w,v)\in E(G)\\ v\in A,\, w\notin A}} f_0(w,v)$$

$$= |f_0|.$$

Now if f_1 is any flow with $|f| < |f_1|$, then we have $|A| < |f_1|$, in contradiction to theorem 6.3.6. $\quad \bullet$

As an immediate consequence of the ideas contained in the proof of the preceding theorem we obtain the following important result.

Theorem 6.3.8. *In each network the magnitude of the maximum flow is equal to that of the minimum cut.* ●

Algorithm 6.3.4 can be used to find a maximum flow in a network. However, difficulties arise concerning the length of the computation.

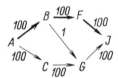

Figure 19.

Let us examine, for example, the network in figure 19 (where the numbers labelling the edges denote the capacities).

It cannot be excluded that in the odd iterations of the cycle consisting of steps 2 to 6 of algorithm 6.3.4 the flow would be augmented along the path A, B, G, J which would lead to augmenting the flow through the edges (A, B), (B, G), and (G, J) by 1, whereas in the even iterations the sequence would be A, C, G, B, F, J which would lead to augmenting the flow through the edges (A, C), (C, G), (B, F) and (F, J) also by 1 and to decreasing the flow through the edge (B, G) by 1. In that case, it would be necessary to execute 200 iterations of the above cycle. Thus, in the worst case, even a small network can be computed for a very long time by algorithm 6.3.4, and admitting irrational values of the capacities, it could happen that the computation would never terminate. On the other hand, with rational capacities it can be proved, at least, that the Ford–Fulkerson algorithm terminates.

Theorem 6.3.9. *If all capacities of a network* **S** *are rational numbers, then algorithm 6.3.4 finds a maximum flow in the network* **S** *in a finite number of steps.*

Proof. We can assume that the capacities are integers because otherwise we can multiply them by the least common multiple of the denominators of all capacities, and we thus obtain a network having integer capacities which algorithm 6.3.4 would process precisely as long as the original network. If all capacities are integers, then the numbers d_i, d_i' and d_i'' and all the values of flow through the edges computed during the execution of algorithm 6.3.4 are also integers. Since the number d in step 3 is always positive, it follows that $d \geq 1$ and hence, each iteration of steps 2 to 6 increases the magnitude of the flow by 1 at least. Therefore, the magnitude of any cut is, due to theorem 6.3.6, an upper bound of the number of iterations of steps 2 to 6. ●

From the reasoning of the preceding proof we immediately conclude that the following statement holds.

Theorem 6.3.10. *In each network with integer capacities there exists a maximal flow which assigns an integer to each edge. Such a flow can be found by means of algorithm 6.3.4.* ●

Improvements of this algorithm which would prevent an overly time-consuming computation have been suggested independently by J Edmonds and R M Karp [92] and E A Dinic [83]. The first method is quite simple: one proceeds just as in algorithm 6.3.4, taking care only that the sequence $v_0, ..., v_k$ generated in step 2 has the least possible length. It can be proved (in a way analogous to the proof of theorem 6.3.9) that such a modified procedure determines a maximum flow in a network of n vertices and m edges in running time $O(nm^2)$, regardless of the size of the capacities of edges.

We shall study the algorithm of Dinic in more detail; it is somewhat more complicated, but it is quicker and, primarily, makes it possible to reduce the problem of searching a maximum flow to the problem of searching a saturated flow in a network of a special type.

The basic scheme of procedure of Dinic's algorithm is the following.

Algorithm 6.3.11. Finding a maximum flow in a network
Input data: a network $S = (G, s, t, c)$.
Task: find a maximum flow f in the network S.
Auxiliary variables: networks S' and S'' with the same set of vertices and the same source and target as S, and q, a flow in the network S''.

1. [*Initiation*] For all edges h of the graph G put $f(h) := 0$.
2. [*Define the network S'*] We define the network S' with the given set of vertices and the given source and target as follows: an ordered pair (v, w) is an edge of S' if, and only if,
    either $(v, w) \in E(G)$ and $c(v, w) - f(v, w) > 0$,
    or, $(w, v) \in E(G)$ and $f(w, v) > 0$.
In the first case, the capacity of the edge (v, w) in S' is the number $c(v, w) - f(v, w)$, and in the latter case, the capacity of (v, w) in S' is $f(w, v)$.

3. ⌊*Define the network* S'⌋ If there does not exist a directed path from the source to the target in the network S'', the computation is terminated and f is a maximum flow of the network S. Otherwise, we define the network S' by deleting from the network S all edges which do not lie on a directed path from the source to the target of S having the least possible length. (The length of a directed path is defined to be the number of edges lying on it.) The edges to be deleted from S are chosen as follows.

(a) For each vertex v we determine the length $\varrho(s, v)$ of the shortest directed path from the source to the vertex v,

(b) for each vertex v we determine the length $\sigma(v, t)$ of the shortest directed path from v to the target t, and

(c) we delete an edge (v, w) from S if, and only if, we have

$$\varrho(s, v) + 1 + \sigma(w, t) > \varrho(s, t).$$

4. [*Find a saturated flow in* S'] Using some of the procedures introduced below, find a saturated flow q in the network S'.

5. [*Augment the flow* f] For each edge (v, w) of the network S' proceed as follows:

if $(v, w) \in E(G)$, put $f(v, w) := f(v, w) + q(v, w)$,

if $(w, v) \in E(G)$, put $f(w, v) := f(w, v) - q(w, v)$.

6. Jump to step 2. ●

Before turning to the properties of algorithm 6.3.11, we prove an auxiliary result.

Lemma 6.3.12. *Let* S_1 *and* S_2 *be two networks defined, in the given ordering, at two successive executions of step 2 of algorithm 6.3.11. Let* $\varrho_i(v, w)$ *for* $i = 1, 2$ *denote the length of the shortest directed path from the vertex* v *to the vertex* w *in the network* S_i, *and in cases where such a path does not exist, put* $\varrho_i(v, w) = \infty$. *Then* $\varrho_1(s, t) \leq \varrho_2(s, t) - 1$.

Proof. Let f_1 and f_2 denote the magnitudes of the flow f at the instant of defining the networks S_1 and S_2, respectively.

The following holds:

if $(v, w) \in E(G)$, then $f_1(v, w) \leq f_2(v, w)$,

and

if $(w, v) \in E(G)$, then $f_1(w, v) \geq f_2(w, v)$.

Consequently, given an edge (v, w) of the network S_2 which is not an edge of the network S_1, then we necessarily have the following:

$$(w, v) \in E(G) \quad \text{and} \quad 0 = f_1(w, v) < f_2(w, v).$$

Therefore, (w, v) must be an edge of the network S_2' created from the network S_1 in step 3.

Now, let $s = v_0, v_1, ..., v_k = t$ be a directed path from the source to the target in the network S_1. We prove by induction that for $i = 1, 2, ..., k$ we have $\varrho_1(s, v_i) \le i$, and in the case that at least one of the edges $(v_0, v_1), (v_1, v_2), ..., (v_{i-1}, v_i)$ does not lie in S_1, then $\varrho_1(s, v_i) \le i - 2$.

The above assertion is trivially true for $i = 0$. Assuming that it holds for $i - 1$ then, if (v_{i-1}, v_i) is an edge of the network S_1, we have

$$\varrho_1(s, v_i) \le \varrho_1(s, v_{i-1}) + \varrho_1(v_{i-1}, v_i) \le i - 1 + 1,$$

and if at least one of the edges $(v_0, v_1), (v_1, v_2), ..., (v_{k-1}, v_k)$ is not an edge of the network S_1, we have

$$\varrho_1(s, v_i) \le (i - 1) - 2 + 1 = i - 2.$$

Further, if (v_{i-1}, v_i) is not an edge of the network S_1, we conclude that (v_i, v_{i-1}) is an edge of the network S_1' and hence, the latter edge lies on the shortest path from the source to the target of the network S_1. The section of the path from the source to the vertex v_i is, then, the shortest path from s to v_i in S_1, and we have $\varrho_1(s, v_i) + 1 = \varrho_1(s, v_{i-1})$. Since, however, $\varrho_1(s, v_{i-1}) \le i - 1$, we obtain $\varrho_1(s, v_i) \le i - 2$.

Next, let $s = v_0, v_1, ..., v_k = t$ be a path of the shortest length from the source to the target in the network S_2. Then we have $\varrho_1(s, t) \le k = \varrho_2(s, t)$. Assuming $\varrho_1(s, t) = \varrho_2(s, t)$, all edges $(v_0, v_1), (v_1, v_2), ..., (v_{k-1}, v_k)$ have to lie in S_1 and hence, the path $v_0, v_1, ..., v_k$ is the shortest path from the source to the target in the network S_1. It follows that all the edges of that path lie also in S_1'. We now use the fact that the flow f_1 which is found in the network S_1' in step 4 of algorithm 6.3.11 is saturated, and hence, for some $i = 1, 2, ..., k$ we have

$$q_1(v_{i-1}, v_i) = c(v_{i-1}, v_i) - f_1(v_{i-1}, v_i)$$

since the expression on the right-hand side is the capacity of the edge (v_{i-1}, v_i) in the network S_1'. We conclude that $f_2(v_{i-1}, v_i) = c(v_{i-1}, v_i)$, see step 5, and therefore the edge (v_{i-1}, v_i) does not lie in the network S_2, in contradiction to the original hypothesis. ●

We will now estimate the speed of algorithm 6.3.11.

Theorem 6.3.13. *Assume that in step 4 of algorithm 6.3.11 we use a procedure which finds a saturated flow in a network of n vertices and m edges in running time at most $h(n, m)$. Then the total running time of the computation required by algorithm 6.3.11 is $O(nh(n, m))$ where n and m are the numbers of vertices and edges of the given network respectively.*

Proof. We first observe that $m \leqq h(n, m)$ because when searching a saturated flow we have to consider the capacity of each edge, and this already requires m steps of computation.

An execution of step 1 takes running time $O(m)$, each execution of steps 2, 3a–3c and 5 takes running time $O(m)$ (see theorem 4.2.2 for the cases 3a and 3b) and one execution of step 6 takes a constant running time. Since the distance between the source and the target of the network S defined in step 2 is at most n and at least 1, it follows from the preceding theorem that steps 2–6 will be executed at most n times. The total running time of the computation is thus

$$O(m) + n[O(m) + O(m) + O(m) + h(n, m) + O(m) + O(1)]$$
$$= O(nh(n, m)). \quad \bullet$$

We will, finally, prove the correctness of algorithm 6.3.11.

Theorem 6.3.14. *Algorithm 6.3.11 determines a maximum flow in a network.*
Proof. We leave to the reader the easy verification that f is a flow throughout the computation. We will just prove the maximality of f at the end of the computation. Denote by A the set consisting of the source s and all vertices into which a directed path leads from s in the network S at the instant of the termination of the computation. The condition of termination implies that A is a cut, and using the same procedure as in the proof of theorem 6.3.7, we can show that the magnitude of the cut A is equal to the magnitude of the flow f at the instant of termination of the computation. By theorem 6.3.6, this implies that the resulting flow is maximum. $\bullet$

Algorithm 6.3.11 transforms the problem of searching a maximum flow in a general network to a repeated performance of the substantially simpler task of searching a saturated flow in a network of a special type which is expressed by the following definition.

Definition 6.3.15. A network is said to be stratified if the set of all edges has a disjoint decomposition into sets $X_0, X_1, ..., X_k$ with the following properties: $X_0 = \{s\}$ and $X_k = \{t\}$, if $v \in X_i$ and $(v, w) \in E(G)$ then $w \in X_{i+1}$, and each vertex different from the source has an incoming edge and each vertex different from the target has an outcoming edge. $\bullet$

Theorem 6.3.16. *The network S' defined in step 3 of algorithm 6.3.11 is stratified.*

Proof. The construction of S' is such that each vertex and each edge of the network lies on a directed path from the source to the target of the network S' with the shortest length [i.e. the length $\varrho(s,t)$]. It is sufficient to put $X_i = \{v \in V(G) | \varrho(s,v) = i\}$. Then, given an edge (v,w) of the network S'', there exists a directed path $s = v_0, v_1, ..., v_k = t$ with $k = \varrho(s,t)$ in the network S_1 and a number i such that $v = v_{i-1}$ and $w = v_i$. It follows that $\varrho(s,v) = i - 1$ and $\varrho(s,w) = i$; consequently, $v \in X_{i-1}$ and $w \in X_i$. The proof of the remaining properties is analogous. ●

Dinic's original method for finding a saturated flow in a stratified network is actually a simplification of the algorithm of Ford and Fulkerson, and it consists of the repetition of the following procedure.

In the given network S we find a path from the source to the target consisting solely of non-saturated edges, i.e. edges h such that $c(h) > f(h)$. (If such a path does not exist, then the given flow f is saturated.) We then augment the flow through all edges of that path by the number d equal to the minimum of the values $c(h) - f(h)$ ranging over all edges lying on the path under consideration.

The special type of a stratified network makes it possible to find and process each path in running time $O(n)$, where n is the number of vertices of the network. Each execution of the above procedure of augmentation of the flow along a given path will saturate at least one edge, and therefore, after at most m repetitions all edges will be saturated, where m denotes the number of edges of the network. The computation will then be terminated. Finding a saturated flow thus requires running time $O(nm)$ and determining a maximum flow in a general network takes running time $O(n^2m)$, which is a better estimate than the estimate $O(nm^2)$ for the algorithm of Edmonds and Karp.

A still quicker algorithm, designed by A V Karzanov [176], makes it possible to find a maximum flow in running time $O(n^3)$. H Hamacher [144] has done some empirical tests with that algorithm. Here we present a simplification of Karzanov's algorithm which is equally quick and which has been described by V M Malhotra *et al* [198].

Algorithm 6.3.17. Finding a saturated flow in a stratified network
Input data: a stratified network $S = (G, s, t, c)$.
Task: find a saturated flow f in the network S.
Auxiliary variables: each vertex and each edge of the graph G can be either open or closed. Furthermore, queues M^+ and M^- containing vertices of the graph G are used, and for each vertex v, the following numbers, $r_{in}(v), r_{out}(v), r(v), g^+(v)$ and $g^-(v)$, are specified. (M^+ will contain those

vertices v for which the flow along all outcoming edges is to be augmented by the value $g^+(v)$, and analogously for M^- and g^-.) Finally, queues R^+ and R^- containing vertices of the graph G are used. (R^+ will contain the vertices to become closed because all their outcoming edges are closed, analogous to R^-.)

1. [*Initiation*] Put $f :=$ the zero flow. Each vertex and each edge is open. Put

$$r_{in}(v) := \sum [c(h) - f(h)],$$

where the sum ranges over all open edges h with the terminal vertex v, and

$$r_{out}(v) := \sum [c(h) - f(h)]$$

where the sum ranges over all open edges h with the initial vertex v.

2. [*Determine the reserves of the vertices*] Put $r(s) := r_{out}(s)$ and $r(t) := r_{in}(t)$, and for each vertex v different from s and t put $r_v := \min (r_{in}(v), r_{out}(v))$. The number $r(v)$ expresses the amount by which the flow through the vertex v can be augmented.

3. [*Determine the reference point*] Choose an open vertex v with the least value of $r(v)$, and denote it by v_0. Put $r_0 = r(v_0)$.
   If $v_0 \neq t$, put $M^+ := \langle v_0 \rangle$; else, $M^+ :=$ empty queue.
   If $v_0 \neq s$, put $M^- := \langle v_0 \rangle$; else, $M^- :=$ empty queue.

4. [*Augment the flow*] The flow from the source to the reference vertex v_0 is augmented by r_0, and the flow from v_0 to the target is also augmented by r_0. The modification of the flow is performed as follows.
   While M^+ is non-empty, repeat step 4a.

4a. Let v be the first vertex in the queue M^+. Choose an open edge (v, w) with the initial vertex v and put $d := \min (g^+(v), c(v, w) - f(v, w))$. Increase the numbers $f(v, w)$ and $g^+(w)$ by the value d, and decrease the numbers $g^+(v), r_{out}(v)$ and $r_{in}(w)$ by the same value d. If $w \neq t$ and w is not in M^+, insert w at the end of M^+, and whenever we now get $g^+(v) = 0$, delete v from M^+. If $c(v, w) = f(v, w)$, execute the following three instructions:
   close the edge (v, w);
   if no open edge has the initial vertex v, insert v into the end of R^+;
   if no open edge has the terminal vertex w, insert w into the end of R^+.
   Analogously, while M^- is non-empty, repeat step 4b.

4b. If v is the first vertex of M^-, choose an open edge (u, v) and put $d := \min (g^-(v), c(u, v) - f(u, v))$. Increase the values of $f(u, v)$ and $g^-(u)$ by d, and decrease the values of $g^-(v), r_{out}(u)$ and $r_{in}(v)$ by d. If $u \neq s$ and u does not lie in M^-, insert u into the end of M^-. If $g^-(v) = 0$, delete v from M^-. If the edge (u, v) has become saturated, close it and, eventually, insert u at the end of R^+ or v at the end of R^-.

5. [*Close edges and vertices*] While R^+ is non-empty, repeat step 5a.

5a. Let v be the first vertex of R^+. Close it and delete it from R^+. Further, close all edges with the terminal vertex v. In cases when this generates vertices with no outcoming open edge, insert such vertices at the end of R^+.

Analogously, while R^- is non-empty, repeat step 5b.

5b. Delete the first vertex from R^-, close it and close also all outcoming edges, and if this generates vertices with no incoming open edge, insert all such vertices at the end of R^-.

6. [*Termination test*] If the source is closed, the computation terminates and f is a saturated flow. Else, go to step 2. ●

We first prove the correctness of algorithm 6.3.17.

Theorem 6.3.18. *At the instant of termination of algorithm 6.3.17, the flow f is saturated.*

Proof. It is not difficult to verify that the above algorithm is designed in such a way that the flow never exceeds the capacities. Moreover, it is clear that if we put $g^+(v) = 0$ for all $v \in M^+$ and $g^-(v) = 0$ for all $v \in M^-$, then each vertex v has the following property at each instant of the computation: the sum of the number $g^+(v)$ with the flow through all the outcoming edges of v is equal to the sum of the number $g^-(v)$ with the flow through all the incoming edges of v. Since at the end of computation the values of g^+ and g^- are all zero, it follows that f is a flow.

We have to prove that at the instant of beginning step 4a there exists an open outcoming edge of the vertex v. This follows from the fact that the total augmentation of the flow through the outcoming edges of the vertex v does not exceed r_0, whereas the reserve for this augmentation is $r_{out}(v)$, and we have $r_0 \leq r(v) \leq r_{out}(v)$. The analogous proof works for step 4b.

We now prove that for each directed path $s = v_0, v_1, \ldots, v_k = t$ the following statement holds at each instant of the computation: if the path contains a closed edge, then it contains a saturated one.

The assertion is clearly true at the beginning of the computation. Now assume that it holds until a certain instant τ at which it is violated. Since saturated edges remain saturated, τ is certainly an instant of closing an edge, say, the edge (v_{i-1}, v_i). If this closing takes place in step 4, then we know that the edge (v_{i-1}, v_i) is saturated, and hence the above assertion still holds. If the closing takes place in step 5, then in the preceding instances either all edges with the terminal vertex v_{i-1} were closed or all edges with the initial vertex v_i were closed. Thus, already before τ there existed a closed edge on the path under consideration, and hence there existed a saturated edge there. This is a contradiction, and thus we have verified that the above assertion cannot be violated at any instant τ.

Since at the instant of the termination all outcoming edges of the source are closed, each path from the source to the target contains a saturated edge, in other words, the flow f is saturated. ●

We now give an estimation of the speed of algorithm 6.3.17.

Theorem 6.3.19. *Algorithm* 6.3.17 *will process a network of n vertices in running time* $O(n^2)$. *In combination with algorithm* 6.3.11 *it will find a maximum flow in a general network in running time* $O(n^3)$.

Proof. By theorem 6.3.13, it is sufficient to prove the first estimate. Since each execution of steps 2 to 6 leads to closing the reference point v_0 which will never be opened again, the number of iterations of those steps is at most n. If m denotes the number of edges of the network, then step 1 can be executed in running time $O(m)$, one execution of step 2 takes time $O(n)$ and the same is true for step 3, and step 6 takes a constant time. Thus, the total running time of steps 1 to 3 and step 6 is

$$O(m) + n[O(n) + O(n) + O(1)] = O(n^2).$$

In step 5 each vertex and each edge is processed at most once throughout the entire computation, and the processing takes a constant time, and hence step 5 requires running time $O(m) = O(n^2)$. The running time of step 4 can be 'put on the bill' of the individual edges and vertices as follows.

If step 4a or 4b leads to a saturation (and hence to closing an edge), then the running time required by this part of the program is put on the bill of the edge under consideration. Each edge is so processed at most once throughout the entire computation. If no edge has been saturated, then a vertex v is deleted either from M^+ or from M^-. The required running time is then put on the bill of the vertex v. During one iteration of steps 2 to 6 each vertex is thus processed at most once in step 4a and at most once in step 4b.

Since each execution of step 4a or step 4b takes a constant time, the bill for the edges is $O(m)$ and that for the vertices is $O(n^2)$. ●

For dense networks, in which $m \sim n^2$, Karzanov's algorithm and the 'three Indians' algorithm 6.3.17 are the quickest methods known today for finding the maximum flow. For sparse networks (e.g. those with $m \sim n$) a series of quicker algorithms has been discovered, see [117, 120, 237, 238, 258]. [238] presents an algorithm with running time $O(nm \log n)$, see also [39, 111, 114, 121, 134, 142, 257]. This is actually Dinic's original algorithm in which the path from the source to the target is not constructed all over again, but the pieces obtained by deleting the saturated edges are stored in a specialised, rather complicated data structure, and the next paths are composed from the stored pieces, which brings a considerable time saving provided that the pieces are sufficiently long.

An important special case of sparse networks is the planar ones. For example, all networks describing road or rail traffic are planar, or almost planar. For such networks there exist algorithms which find the maximum flow both more quickly and more simply than the algorithms designed for general networks.

Especially advantageous are those planar networks which either have the source and target connected by an edge or which make it possible to add such an edge without disturbing the planar drawing. Such networks will be said to be *ST-planar*. It can be proved that each ST-network has a planar drawing such that to the left of the source and to the right of the target there lies neither any vertex, nor any curve representing an edge; see figure 19. It is then clear what is meant by the 'uppermost' directed path from the source to the target; this path is drawn in bold lines in figure 19.

A simple algorithm for finding a maximum flow in an ST-planar network was described independently by C Berge [5] and L R Ford and D R Fulkerson [106].

Algorithm 6.3.20. Finding a maximum flow in an ST-planar network
Input data: an ST-planar network $S = (G, s, t, c)$.
Task: find a maximum flow f in the network S.
  1. [*Initiation*] $f :=$ zero flow.
  2. [*Determine a path*] Find the uppermost directed path

$$s = v_0, v_1, ..., v_k = t$$

from the source to the target in the network S. If such a path does not exist, the flow f is maximum, and the computation is terminated.
  3. [*Augment the flow*] Let d be the minimum of the following numbers

$$c(v_{i-1}, v_i) - f(v_{i-1}, v_i) \qquad \text{for} \quad i = 1, ..., k.$$

Augment the flow through the edges (v_{i-1}, v_i) by the value d.
  4. [*Modify the network*] Delete from the network S all edges (v_{i-1}, v_i) of the path $v_0, v_1, ..., v_k$ for which $c(v_{i-1}, v_i) = f(v_{i-1}, v_i)$, and return to step 2. ●

Theorem 6.3.21. *At the instant of termination of algorithm 6.3.20, f is a maximum flow in the network S.*

Proof. We construct a set A of vertices of the given network S in the following way: in each of the paths considered in step 2 of algorithm 6.3.20 at least one saturated edge is created in step 3. We insert into A all vertices of those paths lying between the source and the initial vertex of the first saturated edge. From geometrical insight it is obvious that A is a cut, and that all edges with initial vertex in A and terminal vertex outside A are saturated. This implies that at the instant of termination of the computation, the magnitude of the flow f is equal to the magnitude of the cut A, and hence, by theorem 6.3.6, f is a maximum flow of the network S. ●

Algorithm 6.3.20 is quite fast in spite of its simplicity.

Theorem 6.3.22. *Algorithm 6.3.20 will process a network of n vertices in running time $O(n^2)$.*

Proof. It is easy to show that one execution of steps 1 to 4 requires time proportionate to the number of vertices or edges of the network. Each execution of steps 2 to 4 will saturate at least one edge and hence the total number of repetitions of those steps is smaller than, or equal to, the number of edges. Finally, it is sufficient to take into consideration that a planar graph of n vertices has at most $6n$ directed edges. ●

The speed of algorithm 6.3.20 can be increased to $O(n \log n)$, see [159]. Let us indicate how this can be achieved.

The path considered in the algorithm descends in the network from above downwards and therefore, for each edge h the computation can be divided into three stages as follows:

the time when the edge h lies under the path,

the time when the edge h lies on the path,

and

the time when the edge h either lies above the path or has been deleted from the network.

Each one of the three stages can be missing, but none is repeated. It is not difficult to verify that the resulting flow through the edge h is equal to the difference $f_2(h) - f_1(h)$, where $f_1(h)$ is the magnitude of the flow at the first instant when h was included in the path under consideration, and $f_2(h)$ is the flow at the instant when h was deleted from the path, or when the computation terminated. It is therefore unnecessary to always modify the flow magnitude of all edges through which something flows: it is sufficient to determine, at every instant, the global magnitude of the flow and the numbers $f_1(h)$ and $f_2(h)$.

Proceeding as above, we face the problem of how to determine the augmentation of flow, and the edges which become saturated and should, therefore, be deleted from the path. It can be shown that exactly those edges h will become saturated for which the value of the number $c(h) + f_1(h)$ is the least possible. Consequently, the augmentation d can be computed according to the following formula: $d = \min\left(c(h) + f_1(h)\right) - |f|$. The edges of the path will thus best be stored in a priority queue according to the values of $c(h) + f_1(h)$. This makes it possible to determine the saturated edges and the augmentation d in running time $O(\log n)$.

The last improvement is achieved by not searching the whole uppermost path all over again, but only the 'detour' forced by deleting the saturated edge.

A nice result of R Hassin [146] shows that algorithm 6.3.20 is a dual version of Dijkstra's algorithm 6.1.2. J Reif [225] described an $O(n \log^2 n)$ algorithm for finding a minimal cut in a general planar network and R Hassin and D B Johnson [147] generalised its result by giving an $O(n \log^2 n)$ planar maximal flow algorithm.

6.4 Maximum matching

We remind the reader of the concept of matching, introduced in §1.2.

The problem of finding the maximum matching, i.e. a matching with the maximum number of edges, of a given undirected graph is another of the tasks for which sufficiently fast algorithms are known; see [90]. Although the problem in itself is seldom encountered, it plays an important role as an auxiliary procedure for the solution of a number of more complicated tasks

The aim of the present section is to describe algorithms determining a maximum matching in an undirected graph. We first study bipartite graphs for which the problem is simpler than in the general case because it is possible to convert it to a search of a maximum flow in a network of a special type.

Let G be a bipartite graph whose vertices are decomposed into two disjoint sets A and B such that each edge has one end-vertex in A and the other one in B. We create a network S as follows: we add to the vertices of G two new vertices, s, the source, and t, the target, and we define the following directed edges of the network S:

(s, a) for all $a \in A$,
(a, b) for all edges $\{a, b\}$ of G with $a \in A$ and $b \in B$,
and
(b, t) for all $b \in B$.

All capacities of edges are equal to 1. The following statement clarifies the relationship between G and S.

Lemma 6.4.1. *There exists a bijective correspondence between integer flows in the network S and matchings of the graph G determined as follows: given an integer flow f in S, then the set of edges $\{\{a, b\} \mid f(a, b) = 1\}$ is a matching of G; conversely, given a matching H of G, the following rule:*

$f(s, a) = 1$ *if an edge of matching H has an end-vertex a,*
$f(a, b) = 1$ *if $\{a, b\}$ is an edge of the matching H,*
$f(b, t) = 1$ *if an edge of the matching H has an end-vertex b,*
$f(h) \quad = 0$ *for all the remaining edges h of S*

defines an integer flow in the network S. The magnitude of the flow f is equal to the number of edges of the matching H.

The proof is obvious because an integer flow through the edges of the network S can have the values $f(h)$ for the individual edges h equal to 1 or 0 only. ●

An example of this correspondence between a bipartite graph and the corresponding network is illustrated by figure 20. The edges of the matching, or those through which the flow is flowing, are drawn bold.

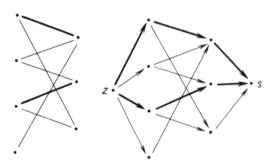

Figure 20.

In the preceding section we showed that in a network with integer capacities there exists a maximum integer flow which, moreover, can be determined by the algorithms of that section. Thus, we can find a matching in a bipartite graph by constructing the above network, applying the algorithms of § 6.3 to find the maximum integer flow through the network, and finding the matching corresponding to the flow. Such a pairing will then be maximum.

The so-called Hungarian method [186] for finding the maximum matching is nothing more than an application of Ford–Fulkerson's algorithm in the described way and formulated without any reference to networks, just in the terms of the bipartite graph. The quickest method for finding a maximum matching known today uses Dinic's algorithm in much the same way. It has been described independently by E A Dinic [84] and J E Hopcroft and R M Karp [151].

Algorithm 6.4.2. Finding a maximum matching of a bipartite graph

We construct a network S from the given bipartite graph in the way described above. We determine a maximum integer flow in it by algorithm 6.3.11, and this yields a maximum matching by the correspondence in lemma 6.4.1. ●

An estimate of the speed of the above procedure is given by the following result.

Theorem 6.4.3. *Algorithm 6.4.2 will process a graph of n vertices and m edges in running time $O(m\sqrt{n})$.*

Proof. We first estimate the duration of one stage of Dinic's algorithm which is shorter than in the general case. Finding a flow-augmenting path requires running time proportionate to the number of edges lying on the path. Since the augmentation of the flow will be an integer, it will be equal to 1 and hence all the edges of the considered path will be deleted from $S''(f)$. The running time needed for deleting one edge is bounded by a constant, and the same is true for edges deleted directly in step 7a without being included in the flow-augmenting path. In one repetition of steps 2 to 6 we then perform activities requiring running time $O(m)$. Thus, it is sufficient to show that the number of stages does not exceed $1 + 2\sqrt{n}$.

Let f_1 and f_2 be integer flows, and let the magnitude of f_2 exceed that of f_1 by k. An edge $\{a, b\}$ of the graph G is said to be of type 1 or type 2 if $f_1(a, b) = 1$ and $f_2(a, b) = 1$, respectively. Each vertex of the graph G lies on at most one edge of type 1, and on at most one edge of type 2.

Let K be the subgraph of the graph G consisting of precisely those edges which are either of type 1, or of type 2. The only type of components of the graph K which contain more edges of type 2 than those of type 1 are the paths $a_1, b_1, a_2, b_2, ..., a_j, b_j$ where each $\{a_i, b_i\}$ is of type 2 and each $\{b_i, a_{i+1}\}$ is of type 1. Each of these components contains one more vertex of type 2 than of type 1. It follows that there exist at least k such paths, and they are pairwise disjoint – consequently none of them has more than

n/k vertices. Moreover, each of those paths yields the following flow-augmenting path:

$$s, a_1, b_1, a_2, b_2, ..., a_j, b_j, t$$

in the network S.

Let T denote the maximum magnitude of a flow in the network S. Clearly, $T \leq n/2$. Now, consider any instant at which the magnitude of the flow created by algorithm 6.3.11 does not exceed $T - \sqrt{n}$. By the above consideration, there exist at least $\sqrt{n}$ paths which are flow-augmenting in the network S. On the shortest of these paths there lie, besides the source and the target, at most $n/\sqrt{n} = \sqrt{n}$ vertices, and hence, the number of stages performed so far is at most $1 + \sqrt{n}$, see lemma 6.3.12.

Since each stage will augment the flow at least by 1, the number of stages needed to increase the maximum from $T - \sqrt{n}$ to T is at most $\sqrt{n}$. Thus, $\sqrt{n}$ of further stages of algorithm 6.3.11 will be sufficient. ●

For general graphs we cannot apply the above translation to flows in networks. Nevertheless, the method of augmenting the matching along a path remains the basic procedure upon which all the hitherto designed algorithms are based.

Definition 6.4.4. A vertex of a graph G is said to be free with respect to the matching H if it is not incident with any edge of the matching H. A path $v_0, v_1, ..., v_k$ is said to be alternating if for all $i = 1, ..., k - 1$, the edge $\{v_{i-1}, v_i\}$ lies in H if, and only if, the edge $\{v_i, v_{i+1}\}$ does not lie in H. A free path is an alternating path both endpoints of which are free.

Theorem 6.4.5. *A matching H of a graph G contains the maximum amount of edges if, and only if, there exists no free path in the graph G with respect to the matching H.*

Proof. If $x_0, x_1, ..., x_{2n+1}$ is a free path with respect to the pairing H, then by deleting from H all the edges $\{x_{2i-1}, x_{2i}\}$ for $i = 1, ..., n$ and adding to H all the edges $\{x_{2i}, x_{2i+1}\}$ for $i = 0, ..., n$ we obtain a matching with more edges.

Conversely, let H and H_1 be two matchings. All components of the graph K containing solely edges which lie either in H or H_1 have one of the following forms: either a cycle of an even length in which the edges of the matchings H and H_1 regularly alternate, or a single edge lying both in H and H_1, or a path with alternating edges from H and H_1. The only form of a component K containing a larger number of edges from the matching H than from the matching H_1 is a path of the last type above in which both of the end edges lie in H_1. Such a path is free with respect to H. Thus, assuming that H is a matching for which there exists a matching H_1, with a larger number of edges, we conclude that a free path with respect to H exists. ●

Theorem 6.4.5, proved by C Berge [60], indicates the following procedure for finding a maximum matching.

Algorithm 6.4.6. A general scheme for finding a maximum matching
Input data: a graph G.
Task: find a matching H in the graph G with the largest number of edges.
1. [*Initiation*] Put H equal to the graph with the same set of vertices as G, but with no edges.
2. [*Improvement along a given path*] While a free vertex v exists in the graph G, repeat the following operations:
search for a free path with respect to H with the initial vertex v;
if such a path $v_0, v_1, ..., v_{2i+1}$ has been found, delete from H the edges $v_1, v_2, ..., v_{2i-1}, v_{2i}$ and add to H the edges $v_0, v_1, ..., v_{2i}, v_{2i+1}$;
if such a path does not exist, delete the vertex v and all incident edges from the graph G and delete v from H. ●

What remains as a problem is finding a free path. The first sufficiently fast algorithm for finding a free path was described by J Edmonds [87]. His algorithm, which belongs to the classical algorithms of graph theory, makes it possible to find a maximum matching in a general graph of n vertices and m edges in running time $O(n^2 m)$, and it is described in a number of textbooks, see e.g. [10]. Algorithms requiring running time $O(nm)$ were presented independently by H N Gabow [113] and E L Lawler [24]. We will describe the former, because the data structures it uses have made it possible to improve the whole procedure further.

For the description of a matching we will use the function $MATE$ determined in such a way that if $\{v, w\}$ is an edge of the matching, then $MATE(v) = w$ and $MATE(w) = v$. The function is undefined at free vertices.

The basic idea of finding a free path is the following. First choose a free vertex v_0. An alternating path with that initial vertex will be called *even* if it has an even number of edges. If $v_0, v_1, ..., v_{2i}$ is an even path, then the edge $\{v_{2i-1}, v_i\}$ lies in the matching. The terminal vertices of even paths will be called *even vertices* (with respect to the given matching and the given vertex v_0). To find a free path with the initial vertex v_0 is the same as to find an even vertex with respect to v_0 which is connected by an edge to a free vertex. For the description of even paths we will use the function $LINK$ defined usually as follows: if $v_0, v_1, ..., v_{2i}$ is an even path which has been constructed by our algorithm, then for all $j = 1, ..., i$ we have $LINK(v_{2j}) =$

v_{2j-2}. Thus *LINK* together with *MATE* make it possible to determine easily an even path terminating in an even vertex: we have

$$v_{2i-1} = MATE(v_{2i}), v_{2i-2} = LINK(v_{2i}),$$
$$v_{2i-3} = MATE(v_{2i-2}), \ldots .$$

The critical instant of all algorithms for finding a free path in a general graph is that in which a cycle of an odd length has been found. Such

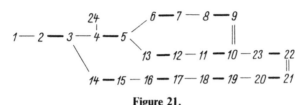

Figure 21.

a situation is illustrated in figure 21 where $v_0 = 1$ and the cycle is $5, \ldots, 13$. In that example, Edmond's algorithm merges the whole cycle to a single point; it searches a maximum matching in the new graph, and then it returns to the original graph. Gabow's procedure makes use of the possibility of a prolongation of the path in the course of searching the even vertices via a certain 'bridge'. For example, in figure 21 the path $1, \ldots, 9$ is prolonged through the edge $\{9, 10\}$ by the vertices $10, 11, 12$ and 13. The function *LINK* is assigned in a different manner to the newly found vertices, and it is chosen to be equal to the bridge under consideration. We thus put $LINK(11) = LINK(13) = \{9, 10\}$ and, analogously, $LINK(6) = LINK(8) = \{9, 10\}$. The function *LINK* defined in this way enables us to find an even path with the initial vertex v_0 as follows: we re-treat from both of the end-vertices of the bridge, and one of the paths thus obtained creates a connection between v_0 and the bridge, whereas a section of the other path yields a connection between the bridge and the terminal vertex of the path. For example, the retreat from the end-vertices 9 and 10 of the bridge in figure 21 yields the paths $1, \ldots, 9$ and $1, \ldots, 5, 13, 12, 11, 10$ from which we can generate even paths to all of the new even vertices $6, 8, 11$ and 13.

The vertices to which the value of *LINK* has not yet been assigned are said to be *unlabelled*, the remaining vertices are *labelled*. A bridge is an edge connecting, at the instant at which it is considered, two labelled vertices. If a bridge $\{x, y\}$ is found, we search the even paths P_x and P_y leading from v_0 to x and y, respectively, which are determined by the functions *MATE* and *LINK*. We determine the place at which they are connected and to each as yet unlabelled vertex lying between the place of connection and x

on the path P_x, or between the place of connection and y on the path P_y, we assign the $LINK$-value $\{x, y\}$ because those are newly discovered even vertices to which a path leads through the bridge $\{x, y\}$.

We illustrate in figure 21 a possibility of speeding up the computation. If we want to find a free path from 1 to 24, we must cross the bridge $21\!=\!22$. But preceding this, we shall evidently cross the bridge $9\!=\!10$, thus determining the vertices as even ones. Therefore, when processing the bridge $21\!=\!22$, we can jump, after the determination of $LINK(23)$, immediately to the next unlabelled vertex 4 of an even path from 1 to 22. We shall thus also specify, for each labelled vertex w, the vertex $FIRST(w)$ as that unlabelled vertex of the even path from v_0 to w which is nearest to the vertex w. For example, after processing the bridge $9\!=\!10$ of figure 21, the value of $FIRST(9)$, originally equal to $MATE(9) = 8$, is changed to 4 and that of $FIRST(10)$ is also changed to 4. If all the vertices of the even path leading to the vertex w have been labelled, we put $FIRST(w) = 0$. When searching the place of connection of two paths leading to the end-vertices of the newly found bridge, we will retreat using the function $FIRST$ instead of $MATE$, which will make it possible to skip the vertices already labelled, and hence, will speed up the computation.

We now describe Gabow's algorithm in detail.

Algorithm 6.4.7. Searching a free path in a graph

Input data: a matching H in the graph G determined by the function $MATE$ and a free vertex v_0 of the graph G with respect to the matching H.

Task: find a free path with respect to H with the initial vertex v_0.

Auxiliary variables: functions $LINK$ and $FIRST$ defined for the vertices of the graph G.

1. [*Initiation*] The function $LINK$ is defined for no vertex w different from v_0. Put $LINK(v_0) := 0$ and $FIRST(v_0) := 0$.

2. [*Choose an edge*] Choose an edge $\{v, w\}$ which has not yet been searched in the direction from v to w and such that $LINK(v)$ is defined. If such an edge does not exist, the computation is terminated; the path we are looking for does not exist.

3. [*Has a free path been found?*] If w is a free vertex different from v_0, go to step 7.

4. [*Choose a continuation*] If $LINK(w)$ is defined, go to step 6; else, continue to step 5.

5. [*Prolong the even path*] Put $z = MATE(w)$. If $LINK(z)$ is undefined, put $LINK(z) := v$ and $FIRST(z) := w$ (else, the edge has brought no new information). Go to step 2.

6. [*Process the bridge* $\{v, w\}$] Execute the following operations 6a to 6g.

6a. [*Initial test*] If $FIRST(v) = FIRST(w)$, jump to step 2.

6b. [*Determine the unlabelled predecessors of v*] Put $r := FIRST(v)$ and $S := r$. While $r \neq 0$ do $r := FIRST(LINK(MATE(r)))$ and $S := S \cup \{r\}$.

6c. [*Determine the connection of the paths*] Put $r := FIRST(w)$, and while $r \notin S$ do $r := FIRST(LINK(MATE(r)))$. If $r \in S$, put $JOIN := r$.

6d. [*Find even vertices between v and JOIN*] Put $r := FIRST(v)$. While $r \neq JOIN$, do $LINK(r) := \{v, w\}$, $FIRST(r) := JOIN$, $r := FIRST(LINK(MATE(r)))$.

6e. [*Find even vertices between w and JOIN*] Put $r := FIRST(w)$. While $r \neq JOIN$, do $LINK(r) := \{v, w\}$, $FIRST(r) := JOIN$, $r := FIRST(LINK(MATE(r)))$.

6f. [*Change FIRST*] For all vertices z for which both $FIRST(z)$ and $LINK(FIRST(z))$ are defined put $FIRST(z) := JOIN$.

6g. [*Go to step 2*]

7. [*Create a new path*] Using the functions $MATE$ and $LINK$ create an even path from v_0 to v and supplement it by the vertex w to a free path with the initial vertex v_0. ●

Since we have described the action of the above algorithm in advance, we do not present a formal proof of its correctness. We shall, however, investigate its speed.

Theorem 6.4.8. *Algorithm 6.4.7 will process a graph of n vertices in running time* $O(n^2)$.

Proof. Step 1 requires running time $O(n)$. Suppose we keep a list (e.g. in the form of a stack) of the vertices v for which the function $LINK$ is defined, but which have the following property: there still exist edges with the initial vertex v which have not yet been searched in step 2 in the direction from v_0. Then step 2 as well as steps 3 and 4 will require running time $O(m) = O(n^2)$, where m is the number of edges of the graph. Steps 5 and 6a require a constant running time, and they are repeated at most m times. If in step 6a we do not jump to step 2, then inside the odd cycle determined by the processed bridge, there is a vertex for which the value of $LINK$ will be determined in steps 6b to 6g. This situation can be repeated at most n times, and one execution of each of steps 6b to 6g takes running time $O(n)$, and therefore, step 6 will altogether require running time $O(n^2)$. Step 7 works in running time $O(n)$. ●

Corollary 6.4.9. *Algorithm 6.4.6 using algorithm 6.4.7 works in running time* $O(n^3)$.

Proof. Each execution of step 2 of algorithm 6.4.6 will decrease the number of free vertices at least by one, and hence, the executions are repeated at most n times. The determination of a free path, or the non-existence of a free path, by algorithm 6.4.7 takes, by the preceding theorem, running time $O(n^2)$, and the other activities of step 2 require running time $O(n)$ in one execution. ●

As we have seen, the part of algorithm 6.4.7 consuming the most running time is step 6. By a careful implementation of that step and, most important, by a suitable choice of data structures, it is possible to decrease the running time to $O(nm)$ for finding a maximum matching.

H Gabow presents in his paper results of practical experiments with the above pair of algorithms executed on an IBM 360/165 computer. Graphs were investigated which had been purposely constructed so as to make the running time proportionate to n^3. For graphs with the number of vertices from 66 to 144, and the number of edges from 968 to 4608, the running times needed for finding a maximal matching were between 0.18 s and 1.6 s. For random graphs (not further specified) of the same sizes the speed was ten times higher.

S Even and O Kariv [98] have combined the procedure of Dinic, Hopcroft and Karp with Gabow's algorithm. The result of that combination is a particularly complicated but fast method which makes it possible to find a maximum matching in a general graph in running time $O(n^2 \sqrt{n})$ or $O(m \sqrt{n} \log n)$, according to the manner of implementation. S Micali and V V Vazirani [207] have simplified the procedure while keeping its speed. Papers [115, 119] deal with fast algorithms for the maximal weighted matching problem.

6.5 Components of graphs

In the present section we study the determination of the degree of connectedness of a graph and the decomposition of the graph into the components of the prescribed degree of connectedness. The importance of the first task has already been mentioned in Chapter 1. The latter one is also very important because it represents the most natural type of decomposition of very large graphs into smaller parts which can then be more easily handled and have specific properties useful for speeding up the computations.

The simplest task is that of determining the connected components of a graph. By theorem 4.2.3, each method of search using algorithm 4.2.1 processes the graph componentwise, and the beginning of the search of a new component is the instant of choosing the new vertex in step 3 of the algorithm. Consequently, the decomposition into connected components can be performed in running time proportionate to the sum of the number of vertices and the number of edges of the graph. Since search is the natural basis for a number of algorithms, the decomposition is often found automatically without requiring special attention.

The problem of finding the articulations of the graph or the 2-connected components is more complicated. We can, naturally, proceed by deleting each vertex individually and then finding out, by the method mentioned above, whether the number of components of the graph has been increased. This naive approach, however, requires running time proportionate to the product $n(n + m)$, where n is the number of vertices and m is the number of edges of the given graph. A more refined procedure has been designed by R E Tarjan [250]. It is based on depth-first searching, and requires a running time of only $O(n + m)$. We proceed by searching the graph and labelling its vertices v by numbers $C(v)$ in the order in which they were reached. In the course of computation we determine, moreover, the number

$$LOWPT(v) = \min \left(\{C(v)\} \cup S_v \right)$$

where

$S_v = \{C(w)|$ there exists a direct or indirect successor z of the vertex v in the tree of depth-first searching for which $\{z, w\}$ is an edge of the graph $G\}$.

The number $LOWPT(v)$ establishes the vertex which we can reach by proceeding from the vertex v in the tree of search and then jumping along an edge of the graph G; if there are more such vertices, we choose that which was searched first. An example of the determination of these numbers

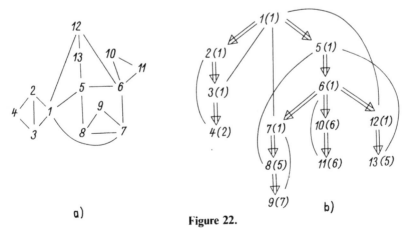

a)   b)

Figure 22.

is illustrated by the parentheses in figure 22b. The vertices in figure 22 are labelled by the numbers $C(v)$. We shall later show a very simple method of determining the numbers $C(v)$.

The method of finding the articulations using the numbers $LOWPT$ is also very simple.

Theorem 6.5.1. *Let a graph* G *be depth-first searched and let the numbers* $LOWPT$ *be determined. Then a vertex* v *is an articulation if, and only if, the following hold:*

either v *is the root of the tree of search, and it has at least two successors in the tree,*

or v *is not the root, and it has a son* w *satisfying* $LOWPT(w) \geq C(v)$.

Proof. Let x be a vertex of the graph G. Denote by $T(x)$ the set consisting of x and all successors of x. It follows from theorem 4.2.7 that a vertex $y \in T(x)$ can only be joined by an edge of the graph G either with the elements of $T(x)$, or with the predecessors of x. In fact, all successors of y lie in $T(x)$, and if y is joined by an edge with its predecessor z, then either z is equal to x, or z is a successor of x, or z is a predecessor of x. In the two former cases we have $z \in T(x)$.

Therefore, assuming that x_1 and x_2 are two sons of the root of the tree of search, then each path from x_1 to x_2 must pass through the root of the tree, and in this case the root is an articulation, by deletion of which the vertices x_1 and x_2 will be separated to distinct connected components. If v is not the root and for a son w we have $LOWPT(w) \geq C(v)$, then since the predecessors of v have smaller labels C than v, it follows that each path going from the vertex w outside $T(w)$ must, after leaving $T(w)$, first pass through the vertex v, and hence v separates the vertex w from the vertices not lying in $T(w)$, e.g. from the father of v.

Let us now prove the converse implication. Let v be a vertex of the graph G and let $w_1, \ldots, w_k$ be its sons. Denote by Q the set of all vertices different from v and all successors of v. By deleting the vertex v from the graph G we obtain a graph in which both Q and each of $T(w_1), \ldots, T(w_n)$ determine a connected subgraph, since each vertex in Q is connected with the root of the tree by a path lying completely inside Q (assuming $Q = \emptyset$). Analogously, each vertex in $T(w_i)$ is connected with w_i by a path lying completely inside $T(w_i)$. Moreover, no pair of the sets $T(w_1), \ldots, T(w_n)$ is directly connected, see theorem 4.2.7. Now, let v be an articulation. Then either v is a root, and hence $Q = \emptyset$ and, necessarily, $k \geq 2$, or v is not a root, and then some of the sets $T(w_i)$ must be a component of the graph obtained from G by deleting the vertex v. The latter means that $LOWPT(w_i) \geq C(v)$ because no element of $T(w_i)$ is connected with a successor of v, and hence an element of $T(w_i)$ can only be connected with the vertex v, or with the vertices in $T(w_i)$, which, however, have the label C larger than that of the vertex v. ●

Using the above characterisation of the articulations we can present an algorithm to find all articulations of a graph.

Algorithm 6.5.2. Finding the articulations of a graph

Input data: a connected, undirected graph G.
Task: establish the set A of all articulations of the graph G.
Auxiliary variables:
  V, the set of all vertices not yet reached,
  E. the set of all edges not yet searched,
  T, the set of all directed edges of the tree of search,
  Z, a stack of vertices of the graph G,
  C, a labelling of the vertices,
  I, the number of vertices that have been reached,
  $LOWPT$, an auxiliary function.
  1. [*Initiation*] $A := \emptyset$, $T := \emptyset$, Z is empty, $V := V(G)$, and $E := E(G)$. Choose a vertex of the graph G, delete it from V, and insert it into the stack Z. Put $I := 1$, $LOWPT(v) := I$, and $C(v) = I$.
  2. [*Search*] Let v be the vertex at the beginning of the stack. Choose an edge $h = \{v, w\}$ lying in E which has an endpoint v. Delete h from E, and if $w \in V$ execute 2a, else execute 2b. If no such edge h exists, execute 2c.
  2a. [*New edge of the tree*] Delete the vertex w from V, insert it into the front of the stack Z, insert the directed edge (v, w) into T, and put $I := I + 1$ and $LOWPT(w) := C(w) := I$.
  2b. [*Transversal of the tree*] Put $LOWPT(v) := \min(LOWPT(v), C(w))$.
  2c. [*Return*] Delete the vertex v from the stack Z. If Z is still non-empty, and if z is the vertex at the front of Z. put

$$LOWPT(z) := \min(LOWPT(z), LOWPT(v)).$$

  3. [*Continuation test*] If Z is non-empty, go to step 2.
  4. [*Find the articulations*] For all edges $(v, w) \in T$ such that $1 < C(v) \le LOWPT(w)$ put $A := A \cup \{v\}$. If, in the course of searching T, it has turned out that the root of the tree of search (i.e. the vertex v with $C(v) = 1$) is an initial vertex of two edges of T, insert v into A too. ●

In order to prove the correctness of algorithm 6.5.2, it is sufficient to show that the algorithm correctly establishes the values $LOWPT$, which we leave to the reader as an exercise. The rest then follows from theorem 6.5.1.

The following theorem shows that the above algorithm finds the articulations very quickly.

Theorem 6.5.3. *Algorithm 6.5.2 will find all articulations of a connected graph of m edges in running time O(m).*

   Proof. A connected graph has at most $m + 1$ vertices. Steps 1 to 3 of the above algorithm represent the depth-first search with an additional execution of several instructions for each edge. They thus require running time $O(m)$. Step 4 requires running time proportionate to the number of elements of the set T which is also $O(m)$.   ●

   J E Hopcroft and R E Tarjan [153] proved that the depth-first search also makes it possible to find the biarticulations and the 3-connected components of a graph in running time proportionate to the sum of the number of edges and the number of vertices of the graph. The algorithm performs an analysis of the graph in precisely the same way in which it is performed by the procedure for searching a planar drawing of a graph (due to the same authors) which will be described in §6.7 below. It differs from the latter procedure only in the final processing of the information obtained.

   In contrast to the problem of finding the articulations and the biarticulations of a graph, the investigation of the n-connectedness of a graph for $n = 4$ is very time-consuming. Using the determination of a flow in a related network, it is possible, for two given vertices u and v of a graph, to determine in running time $O(n^3)$ the minimum size of a set A, the deletion of which separates u and v into distinct connected components of the graph. If this is performed for all pairs of vertices, we find the degree of connectedness of the graph as the minimum of the sizes of the separating sets in running time $O(n^5)$. S Even and J E Hopcroft [97] and S Even and R E Tarjan [99] have described an improvement of the above method which, using Dinic's algorithm, works in running time $O(m^2 n)$, where n is the number of vertices and m is the number of edges. We see that the last algorithm is also not very fast since for $m \sim n^2$ we obtain the estimate $O(n^{4.5})$. Recently D Matula [203] showed how to determine edge connectivity of a graph in $O(nm)$ time. Some other papers related to connectivity are [167, 193, 203, 208].

   We finally turn to the problem of finding the strong components of a directed graph. A method of finding the strong components in running time proportionate to the sum of the number of vertices and edges has been described by R E Tarjan [250]. His procedure is also based on depth-first searching, in this case using algorithm 4.2.9. It can be proved that in each strong component of a depth-first searched directed graph there lies a vertex, called the *root* of the strong component, which determines the remaining vertices of that component as follows: a vertex w is an element of a strong component Q with the root v if, and only if, either $v = w$, or w is a successor of v in the forest of search and on the path from v to w in the forest there does not lie any other root of any of the strong components.

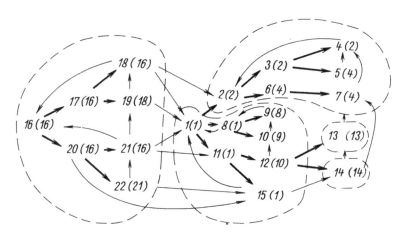

Figure 23.

The whole situation is illustrated by figure 23 where the roots of the strong components are the vertices 1, 2, 13, 14 and 16. The strong components there are denoted by dashed lines, and the edges of the forest of search are denoted by bold lines.

Thus, the roots determine the strong components in a way similar to that in which the articulations determine the 2-connected components in the course of the search by the procedure described at the beginning of this section. The analogy goes even further since the roots can be determined by a function *LOWLINK* (indicated in figure 23 by parentheses) defined in a manner analogous to *LOWPT*:

$$LOWLINK(v) = \min\left(\{C(v)\} \cup S'_v\right),$$

where S'_v is the set of the numbers $C(w)$ for all vertices w with the following properties:

(a) either $v = w$, or there exists a vertex z which is a successor of v in the tree of search and such that (z, w) is a directed edge of the graph G;

(b) there exists a vertex u such that u and w lie in the same strong component of the graph G and both v and w are successors of the vertex u in the tree of search.

It can be proved that v is a-root of a strong component if, and only if, we have $LOWLINK(v) = v$, see figure 23. The proof of that statement is more difficult than that of 6.5.1, and we thus do not present it here. Also the determination of the values of the function $LOWLINK$ is more complicated than in the case of the function $LOWPT$, although the difficulty lies rather in understanding the principle of the corresponding algorithm than in the number of instructions of the corresponding program, or the speed of the resulting computation. The roots of strong components must be determined already in the course of search in order to make it possible to verify, for other vertices, the validity of the second condition in the definition of $LOWLINK$.

6.6 Isomorphism of trees

In this section we present algorithms by means of which it is possible to determine very quickly whether two given trees are isomorphic or not. We will first study rooted trees, and then we will apply the described algorithm to the search of isomorphism of undirected trees. In both cases the algorithms are going to be applicable to vertex-evaluated trees as well, and the generalisation to edge-labelled trees is also easy.

The problem of determining the isomorphism of two trees is important in its own right, but it is most often encountered when one investigates isomorphisms of more complicated objects for which there is a decomposition into parts which are interrelated by a structure that can be described by a tree, or a rooted tree.

The basic algorithm for vertex-labelled rooted trees can be described as follows.

Algorithm 6.6.1. Isomorphism of rooted trees
Input data: vertex-labelled rooted trees T_1 and T_2.
Task: decide whether there exists an isomorphism of the given rooted trees preserving the given labelling of vertices.
Auxiliary variables: the labelling of a vertex v is denoted by $Q(v)$. For each vertex we use a number $C(v)$ and a list of numbers $L(v)$.

1. [*Determine the number of vertices and the depth*] For both trees determine the number of vertices and the depth, as well as the number of vertices on all of the individual levels. All the corresponding numbers obtained for the two trees must be equal, otherwise the trees cannot be iso-

morphic. Let h denote the depth and let n_j be the number of vertices on the jth level and n the total number of vertices.

2. [*Label by the numbers C*] For $j := h, h - 1, ..., 1$ do steps 2a to 2d.

2a. [*Create L*] For each vertex v of the jth level of the tree T_1 or T_2 we create a list $L(v)$ as follows: after the label $Q(v)$ of v we insert, if v has sons, the sequence of numbers $C(v_1), ..., C(v_k)$, where $v_1, ..., v_k$ are the sons of v ordered in such a way that the sequence $C(v_1), ..., C(v_k)$ is non-decreasing.

2b. [*Order the level*] For $i = 1$ and 2 order the jth level of the tree T_i into a non-decreasing sequence in lexicographic order of the values of the lists $L(v)$. The resulting sequence is denoted by $v_{i1}, v_{i2}, ..., v_{ik}$ for $k = n_j$.

2c. [*Compare the levels*] If there exists $p = 1, ..., k$ such that the list $L(v_{1p})$ is different from the list $L(v_{2p})$, then T_1 and T_2 are non-isomorphic.

2d. [*Label the level*] Define the numbers C for the vertices of the jth level as follows:

$C(v_{i1}) := 1,$
if $L(v_{ip}) = L(v_{i,p-1})$ then $C(v_{ip}) := C(v_{i,p-1}),$
else $C(v_{ip}) := C(v_{i,p-1}) + 1.$

3. [*Terminate*] If the algorithm has not stopped during the execution of step 1 or step 2c, then the trees T_1 and T_2 are isomorphic. ●

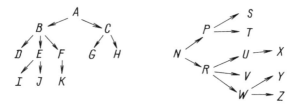

Figure 24.

We first illustrate the determination of isomorphism of trees in figure 24. Assume that the label of the vertices H and T is 2, and the remaining labels are 1. For each vertex of the fourth level we have $L = (1)$, and hence, $C = 1$. Further, we obtain

$$L(D) = L(G) = L(S) = L(V) = 1,$$
$$L(F) = F(U) = (1, 1),$$
$$L(E) = L(W) = (1, 1, 1),$$
$$L(H) = L(T) = (2),$$

and the ordering of the third level is D, G, F, E, H in the tree T_1 and S, V, U, W, T in the tree T_2. Next, we have

$$C(D) = C(G) = C(S) = C(V) = 1,$$
$$C(F) = C(U) = 2,$$
$$C(E) = C(W) = 3,$$
$$C(H) = C(T) = 4,$$
$$L(B) = L(R) = (1, 1, 2, 3),$$
$$L(C) = L(P) = (1, 1, 4)$$

and the ordering of the second level is B, C in the tree T_1 and R, P in the tree T_2. Finally,

$$C(B) = C(R) = 1,$$
$$C(C) = C(P) = 2,$$
$$L(A) = L(N) = (1, 1, 2),$$
$$C(A) = C(N) = 1.$$

Algorithm 6.6.1 turns both of the given trees into ordered rooted trees whose levels are ordered in step 2b. If we need information about the concrete rule of the isomorphism, we execute depth-first (or breadth-first) search of both trees. If we proceed simultaneously in both trees, then the vertices entered at the same instant must correspond under the isomorphism since we have seen in §4.1 that the depth-first (or breadth-first) search of an ordered rooted tree is uniquely determined.

Theorem 6.6.2. *Algorithm 6.6.1 correctly decides whether the trees T_1 and T_2 are isomorphic or not.*

Proof. The equality of the depths and of the numbers of vertices on the individual levels is a necessary condition for the existence of an isomorphism. We now prove by induction that two vertices v and w of the jth level of the tree T_1 (or T_2) will obtain in step 2 the same value of the number C if, and only if, the labelled subtrees determined by the vertices and all their respective successors are isomorphic (with preservation of the labelling). The equality $C(v) = C(w)$ takes place if, and only if, we have $L(v) = L(w)$. If the last equality holds, then $Q(v) = Q(w)$, and moreover, the two lists above have the same length. Consequently, v and w have the same number of sons. The assertion we are proving thus holds for the leaves. If v and w have sons, then by induction hypothesis the equality $L(v) = L(w)$ implies that when comparing the sons in step 2a and when creating the lists $L(v)$ and $L(w)$, the labelled subtrees determined by the corresponding sons of the vertices v and w are isomorphic. Consequently, the labelled subtrees determined by v and w are isomorphic (with preservation of the evaluation).

If the computation reaches step 3, the roots of the trees T_1 and T_2 have the same value of the number C and hence the evaluated trees T_1 and T_2 are isomorphic. ●

When investigating non-labelled trees, we can represent each vertex as a labelled vertex with the value 1, and then use the above algorithm. In this case the elements of the list L and the numbers C are integers between 1 and the number of vertices of the trees, and consequently, for the ordering in step 2 we can use algorithm 5.6.6 which works in running time $O(n)$. The same is true for labelled trees such that the labels Q are integers making it possible to use sorting by distribution.

Theorem 6.6.3. *Algorithm 6.6.1 will process generally labelled rooted trees in running time $O(n \log n)$ where n is the number of vertices of the trees. When applied to unlabelled rooted trees, and when using algorithm 5.6.6 for the determination of the ordering in step 2b, it requires running time $O(n)$.*

Proof. The distribution of the vertices to levels can be performed in running time $O(n)$ by applying the search, and thus step 1 requires running time $O(n)$. If the jth level has n_j vertices then the sum of the lengths of the lists L in the jth level is $n_j + n_{j+1}$ (or n_h in the last level). The sum of the lengths of the lists is decisive for the speed of algorithm 5.6.6, and hence, we see that step 2a requires running time $O(n_j + n_{j+1})$. The same is true, in the unlabelled case, for step 2b. Steps 2a and 2d also require running time $O(n_j + n_{j+1})$. If we cannot use search by distribution in step 2b, the required running time will be $O(N_j \log N_j)$, where $N_j = n_j + n_{j+1}$. The total running time of one execution of step 2 is, therefore, either

$$O(n_1 + 2n_2 + \dots + 2n_h) = O(n),$$

or

$$O((n_1 + 2n_2 + \dots + 2n_h) \log n) = O(n \log n).$$

Step 3 is a comment only. ●

When dealing with edge-labelled rooted trees, we can rewrite the value of each edge as the value of its end-vertex, and then apply algorithm 6.6.1 again.

The assertion about the existence of a centre or a bicentre of an indirected tree presented in Chapter 1 yields also a possibility of applying algorithm 6.6.1 to the case of undirected trees.

Algorithm 6.6.4. Isomorphism of trees

Input data: vertex-labelled undirected trees T_1 and T_2.

Task: decide whether there exists an isomorphism of the given trees preserving the given labelling of vertices.

1. [*Determine the centres*] In both of the given trees determine the centre or bicentre. If one of the trees has a centre and the other one a bicentre, then they are not isomorphic. If both have a bicentre, we add, in each of the trees, a new vertex inside the edge determining the bicentre, and we label the new vertex by 1. This creates new trees which both have a centre, namely, the new vertex, and thus this translates the situation to that of two trees with a centre. Those new trees are isomorphic if, and only if, the given trees T_1 and T_2 are isomorphic.

2. [*Orient the trees*] Both of the given trees are oriented in the direction from the centre.

3. [*Determine the isomorphism*] The given labelled trees T_1 and T_2 are isomorphic (with preservation of the labelling) if, and only if, the labelled rooted trees created in step 2 are isomorphic (with preservation of the labelling). Verify the latter by applying algorithm 6.6.1. ●

The determination of orientations in step 2 can be performed by any type of search which starts at the centre of the tree. The determination of the centre can be performed by the following algorithm which is analogous to algorithm 6.1.7 searching a topological ordering of a graph.

Algorithm 6.6.5. Searching for a centre or a bicentre of a tree

Input data: a tree T.

Task: find the centre or the bicentre of the tree T.

Auxiliary variables: number N denoting the number of remaining vertices of the tree, lists $L1$ and $L2$ of vertices of the tree T, and the number $D(v)$ which, for each vertex v, is equal to the degree of v with respect to remaining vertices.

1. [*Initiation*] Put $D(v) := 0$ for each vertex v. The lists $L1$ and $L2$ are empty, and N is the number of all vertices of the tree T.

2. [*Determine the degrees*] For each edge $\{u, v\}$ of the tree T put $D_u := D_u + 1$ and $D_v := D_v + 1$. (This determines the degrees of the vertices in T.)

3. [*Create L1*] Store into $L1$ all vertices v such that $D_v \leq 1$.

4. [*Termination test*] If $N = 1$, the computation is terminated: the list $L1$ contains a single vertex which is the centre of the tree T. If $N = 2$, the computation is also terminated: the list $L1$ contains two vertices, the bicentre of the tree T.

5. [*Empty L1*] While *L1* is non-empty, perform the following operations.

  5a. [*Choose a vertex*] Choose a vertex v_0 in the list *L1*.

  5b. [*Decrease D_v*] The vertex *w* which lies in *T* and is a neighbour of v_0 will be subjected to the following instruction: $D_w := D_w - 1$, and if $D_w = 1$, insert *w* into *L2*.

  5c. [*Delete v_1*] Delete the vertex v_0 and the edge incident with v_0 from the tree *T*. Delete v_0 from *L1* and decrease *N* by 1.

  6. [*Terminate the deleting stage*] Put $L1 := L2$. Empty *L2* and proceed to step 4. ●

Theorem 6.6.6. *Algorithm 6.6.5 will find the centre or the bicentre of a tree of n vertices in running time $O(n)$.*

Proof. The number of edges of the tree is $n - 1$. Steps 1 to 3 thus take running time $O(n)$, and each performance of steps 4 to 6 deletes at least two vertices from *T*. Consequently, the latter steps will be repeated at most $n/2$ times (not counting the final performance of step 4). One execution of steps 4 to 6 takes a constant running time, and hence, those steps take altogether running time $O(n)$. One execution of steps 5a, b, c requires running time proportionate to the number of neighbours of the vertex v_0, and hence, step 5 takes running time proportionate to the sum of the degrees, which is $O(n - 1)$. Consequently, step 5 also works in running time $O(n)$. ●

The above result yields the following estimate.

Theorem 6.6.7. *Algorithm 6.6.4 which in step 1 uses algorithm 6.6.5 and in step 3 uses algorithm 6.6.1 will determine the isomorphism of two trees of n vertices in running time $O(n)$, and the isomorphism of generally vertex-labelled trees in running time $O(n \log n)$.* ●

6.7 Planarity testing of graphs

Searching for a planar drawing of a given graph is another problem for which fast algorithms are known. In the present section we describe two of them. They both proceed by first finding a cycle of the given graph, and after drawing the cycle in the plane they successively add further parts of the graph until they either produce a planar drawing of the graph, or prove its non-existence.

In order to describe those algorithms we will need several new concepts.

Definition 6.7.1. Let G be a graph and H a subgraph. By a chord of the graph G with respect to H is meant a path in the graph G both end-vertices of which lie in H, whereas no inner vertex nor any edge of the path lie in H.

By a part of the graph G with respect to H is meant either an edge which lies outside H although both of its end-vertices lie in H, or a connected component C of the graph $G - H$ to which we add all edges with one end-vertex in C and the other in H, and all end-vertices of the added edges lying in H. (By $G - H$ we mean the graph obtained from G by deleting all vertices of the graph H and all edges incident with the deleted vertices.)

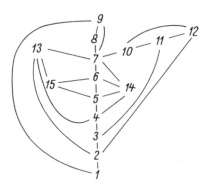

Figure 25.

For example, denote by H the subgraph of the graph in figure 25 consisting of the cycle $1, \dots, 9$. Then the following paths are examples of chords: $7, 9$; $7, 13, 15, 5$; $7, 10, 11, 12, 2$. The parts are determined by the sets of vertices $\{10, 11, 12\}$, $\{14\}$, and $\{13, 15\}$ in the sense that the part consists of all edges mutually connecting the vertices of the set under consideration, or connecting those vertices with the vertices of the cycle $1, \dots, 9$. The edge $\{7, 9\}$ is also a part.

The first algorithm which we present is based on the idea of L Auslander *et al* [49, 135] and it has the following basic scheme.

Algorithm 6.7.2. Finding a planar drawing

Input data: a graph G_0.

Task: find a planar drawing of the given graph.

1. [*Check the number of edges*] If the number of edges of the given graph is bigger than $3n - 3$ where n is the number of vertices, then the graph is not planar, and the computation is terminated.

2. [*Decompose to components*] Decompose the graph G_0 into 2-connected components. For each component, denoted further by G, execute steps 3 and 4.

3. [*Find a cycle*] Determine a cycle K in the graph G. Choose a drawing r of K in the plane, and put $H := K$.

4. [*Process the parts*] For each part C of the graph G with respect to the cycle K execute the following operations 4a and 4b.

4a. [*Determine the planarity of the part*] Determine whether the graph obtained from the connection of the part C and the cycle K is planar. If the answer is negative, then the computation is terminated because neither the graph G nor the graph G_0 is planar.

4b. [*Add the part*] At this instant, H consists of the cycle K and those parts of the graph G with respect to K which have already been searched, and r is a planar drawing of H. By adding to H the part C currently processed, we obtain a graph H_1. We then have exactly one of the following possibilities.

(i) The drawing r can be extended to the graph H_1 using the drawing found in step 4a. In that case put $H := H_1$ and extend r to H_1.

(ii) The drawing r cannot be extended to H_1, but it can be modified so as to enable such an extension. By a modification is meant the re-drawing of some of the parts of the component G with respect to K from the unbounded face given by the circle K into a bounded one, or vice versa. If such a modification is possible, put $H := H_1$ and let the new value of r be the modification of the original one, extended to H_1.

(iii) The drawing r cannot be extended to H_1, nor can its modification described above be extended to H_1. Then the computation is terminated, and neither G nor G_0 is a planar graph.

5. [*Connect the results obtained for the components*] From the planar drawings constructed of the 2-connected components G of the graph G_0 create a planar drawing of the whole graph G_0. ●

The above scheme is illustrated in figure 26a. Assume that the parts $b, b_1, ..., b_4$ have already been drawn. If it is our goal to extend the drawing to part a, then part b prevents it from being drawn outside the main cycle. Deciding to draw part a inside the cycle we have to turn part b_1 from the inside of the cycle to the outside and this forces us to turn parts b_2, b_3 and b_4 too. Another possibility is to turn part b from the outside of the cycle inside, and then part a can be drawn outside. If part b_5 has already been drawn, then none of the above modifications are possible, from which we would conclude that the graph in figure 26a does not have a planar drawing.

Algorithm 6.7.2 has been described above very roughly, and in order to obtain the algorithm of J Hopcroft and R E Tarjan [155] which works in running time $O(n)$, we would need quite refined execution of the individual steps of the algorithm. We therefore omit a detailed description of the proce-

Combinatorial Algorithms

dure, and we mention only briefly some of the basic ideas of implementation. All details can be found in the above-mentioned paper where a full description in Algol W is presented (see also [80] for small corrections). The book [34] also presents a detailed description.

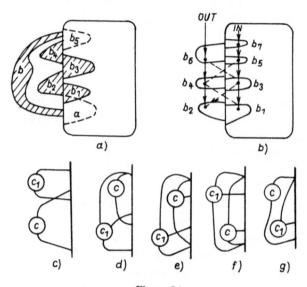

Figure 26.

J Hopcroft and R E Tarjan have shown that the key to a fast execution of computation according to algorithm 6.7.2 is depth-first searching. This must be done in such a way that the following conditions (i) to (v) be fulfilled.

(i) A vertex which we do not enter in the course of search before the second run must be the initial vertex. (In figure 25 this means that we start, for example, in 1, then we pass from 2 to 9, and finally we pass through the edge leading from 9 to 1.)

The first part of a search with the property (i) will determine a cycle which is going to be chosen as K in step 3 of the above algorithm. Assume that its vertices are numbered 1 to k. By the point of contact of an arbitrary part C of the graph G with respect to K is understood any vertex lying in C as well as in K. The point of contact which has the highest (or lowest, or second lowest) enumeration on the cycle will be called the upper (the lower, or the second lower, respectively) point of contact. Further conditions imposed on the search are the following.

(ii) If we enter a part of the graph G with respect to K, then the edge through which we are entering has the upper point of contact as the initial vertex and we are not going to leave the part until all vertices and all edges of the part have been searched. (In figure 25 the part determined by the vertices 13 and 15 must be entered through the vertex $(7, 13)$.)

(iii) After having entered the part C we first meet the cycle K in the lower point of contact of C and K. (In figure 25 we use, after entering the part with the vertices 13 and 15, the edge $(13, 2)$ before each of the edges $(13, 14)$, $(15, 5)$ or $(15, 6)$.)

(iv) Let C_1 and C_2 be two parts of the graph G with respect to K, and let x_i, y_i and z_i, for $i = 1$ or 2, denote the enumerations of the upper, lower and second lower (respectively) point of contact of the part C_i and the cycle K. Assuming that one of the following conditions is fulfilled: $x_1 > x_2$, $x_1 = x_2$ and $y_1 < y_2$, or $x_1 = x_2 = z_1$, $y_1 = y_2$, and $z_2 < x_2$, then we enter the part C_1 before the part C_2. (In figures 26c, d we enter the part c_1 first, in figure 26g we enter c first, and in figures 26e, f the order is not determined.)

(v) Let C_1 and C_2 be parts of the graph G with respect to K such that C_1 was searched before C_2. Then C_1 and C_2 cannot be drawn into the same face determined by the cycle K if, and only if, there exists a point of contact of the part C_1 whose enumeration on the cycle K is smaller than that of the upper point of contact of the part C_2, and larger than that of the lower point of contact of C_2. (See figures 26c to 26f.)

Condition (v) follows from condition (iv) because figure 26 illustrates all possible interrelations of parts which are 'obstacles' for each other, and the only possibility of failure of condition (v) is in the case illustrated in figure 26g. This would occur if part c_1 were drawn before part c, and this is excluded by condition (iv) of course. Due to condition (v) it is possible to simply determine in step 4 of algorithm 6.7.2 whether or not the constructed planar drawing can be extended to the part currently added, or eventually, which parts must be turned to make such an extension possible.

In order to fulfil conditions (i) to (v), it is necessary to use an auxiliary depth-first search at the beginning of computation, before the main search, the aim of which is to construct the following functions:

$$LOWPT1(v) = \min\left(\{C(v)\} \cup S_v\right),$$
$$LOWPT2(v) = \min\left(\{C(v)\} \cup (S_v - \{LOWPT1(v)\})\right).$$

Here $C(v)$ denotes the enumeration of the vertices in the order in which they are reached by the auxiliary search, and S_v is the set defined via the tree of the auxiliary search in the same way as in the definition of the function $LOWPT$ in § 6.5. The function $LOWPT1$ is identical with the function

LOWPT of §6.5 where its computation in running time $O(n)$ has been described. The computation of the function *LOWPT2* is analogous.

Using the above functions *LOWPT1* and *LOWPT2*, we determine the following numbers $F(h)$ where $h = (v, w)$ is a directed edge of the graph G:

$$F(h) = \begin{cases} C(w) & \text{if } (v, w) \text{ is not an edge of the tree of the auxiliary search,} \\ LOWPT1(w) & \text{if the first case does not occur and} \\ & LOWPT2(w) < C(v) \\ LOWPT2(w) + \frac{1}{2} & \text{if the first two cases do not occur.} \end{cases}$$

For the second main search we use the same initial vertex as for the first one, and for further procedure we always choose an edge h with the smallest possible value of $F(h)$.

For example, if we search the graph in figure 25 in the order given by the enumeration of the vertices, the functions *LOWPT1* and *LOWPT2* will take the following values:

	1	2	3	4	5	6	7	8	9	10	11	12	13	14	15
LOWPT1	1	1	1	1	1	1	1	1	1	2	2	2	2	4	5
LOWPT2	1	2	2	2	2	2	2	7	7	3	3	10	4	5	6

The directed edges with the initial vertex 7 have the following values of F:

	(7, 6)	(7, 8)	(7, 9)	(7, 10)	(7, 13)	(7, 14)
F	6	1.5	9	2	2	4

and, for example, the edge $(6, 15)$ is assigned the value 5.5.

As a consequence of the procedure described above we see that one first enters the vertices with low values of *LOWPT*, and this guarantees that conditions (i) to (v) above will be satisfied.

We have shown how to proceed in order to determine which parts are mutual obstacles as simply, and hence, as quickly, as possible. Another stage of the above algorithm requiring a careful performance is the modification of the drawing r in step 4b(ii). If the addition of some of the parts requires turning some previously drawn part into a different face, then it is often necessary to turn whole 'blocks' composed of a lot of parts. This modification could be rather time-consuming. It is, however, possible to prove that a search subject to the above conditions (i) to (v) will process the parts which belong to the same block in direct succession, and thus makes it possible to identify the blocks quite easily. Once the blocks have been specified, then turning them over into a different face can be performed very quickly by means of the following data structure. For each part we store the shape of its drawing inside the cycle K. This information is not going to be changed in the course of the computation. Furthermore, we create linked lists *IN* and *OUT* into which we store the parts which are to be drawn inside or outside the cycle, respectively, in the order in which they are processed.

We present an example in figure 26b. When a block is to be turned, we perform it by simply switching the pointers between the lists *IN* and *OUT*. For example, if in figure 26b it is necessary to turn the block consisting of the parts b_3 and b_4, we perform the switching of pointers as indicated by the dashed lines.

The algorithm of Hopcroft and Tarjan works in running time $O(n)$, but it is very complicated, which causes difficulties of programming, and moreover, its constant of linear proportion is relatively large. Therefore, for smaller graphs the following procedure of G Demoucron *et al* [79] may be more suitable.

Algorithm 6.7.3. Finding a planar drawing

Input data: a graph G.

Task: find a planar drawing of the graph G.

Auxiliary variables: a subgraph H of G already drawn by a planar drawing r, and a chord p of the graph G with respect to H.

1. [*Determine a cycle*] Determine a cycle H in the graph G and choose its planar drawing r.

2. [*Termination test*] If $G = H$, the computation terminates: G is a planar graph with a planar drawing r.

3. [*Determine the faces for drawing the parts*] Let $C_1, ..., C_n$ be parts of the graph G with respect to H, and let $S_1, ..., S_k$ be faces obtained from the drawing r of the graph H. We say that the part C_i can be placed in the face S_j if the planar drawing r can be extended to the graph obtained by connecting the part C_i with the subgraph H to a planar drawing which draws each vertex of C_i inside the face S_j of the original graph H, including the circumference of the face S_j. One of the following cases applies.

3a. [*Non-planarity proof*] Some of the parts C_i cannot be placed in any face. Then computation is terminated: the graph G is not planar.

3b. [*Uniquely placed part*] There exists a part C_i which can be placed in one face only, say, the face S_j. Put $C := C_i$ and $S := S_j$, and go to step 4.

3c. [*Non-uniquely placed parts*] Each part C_i can be placed in two different faces at least. Choose an arbitrary part C, and let S be any face in which C can be placed. Go to step 4.

4. [*Add a chord*] Choose a chord of the graph G contained in the part C. Add the chord to the graph H, and extend the drawing r to the extended graph H in such a way that the chord is drawn in the face S. Go to step 2. ●

Algorithm 6.7.3 extends the graph *H* carefully: instead of drawing the whole part *C* in the face *S* in step 4, it only draws a chord lying in the part. Due to this feature, it is not necessary to modify the drawing *r*, in contrast to algorithm 6.7.2. This makes the present algorithm simpler. The running time estimate of algorithm 6.7.3 in the worst case is $O(n^2)$ since it can happen that for the addition of a chord in step 4 it is necessary to check all the parts, the number of which càn be quite high. Practical experience shows, however, that the above case 3b is rather frequent, and therefore the algorithm is usually quite fast.

The choice of the face *S* for the drawing of a chord was forced in case 3b. At first sight, it might seem that the choice of face in case 3c can lead to a planar drawing which cannot be extended to all of the graph *G*, even though *G* is planar. The authors of the algorithm have proved, however, that such a situation cannot occur and hence that the algorithm is correct. Although the idea of the proof is quite simple, it requires considering a large number of possible cases, and we therefore do not present the proof here.

We first briefly mention the problem of determination of the faces in which a given part *C* of the graph *G* with respect to *H* can be placed. Denote by $D(C)$ the set of all points of contact of the part *C*, i.e. the vertices which lie in both *C* and *H*. Let $I(v)$ denote the set of all faces which contain the vertex *v* in their circumference. Then the set of all faces in which the part *C* can be placed is the following intersection:

$$\bigcap_{v \in D(C)} I(v)$$

The determination of the above intersection is easy, and so is the actualisation of the values of $D(v)$ and $I(v)$ in step 4 ensuing from the fact that the newly added chord will divide the face *S* into two.

F Rubin [228] compared Tarjan's implementation of algorithm 6.7.2 in Algol W with his own improved implementation of algorithm 6.7.3 in PL/I (his paper contains a very detailed description of the latter). The algorithms were examined on an IBM 360/67 computer, the inputs of which were randomly generated planar graphs with the maximum number of edges. Whereas R Tarjan states for the running time the empirical formula $t \doteq 12\,500n - 70\,000\,\mu s$, Rubin's implementation required $t \doteq 0.005n^2 + 6087n - 20\,797\,\mu s$. This yields $t = 12.5\,s$ (Hopcroft–Tarjan) and $t = 6\,s$ (Rubin) in the case of a graph of 1000 vertices. Although such comparisons must always be taken with a lot of reserve, the above results indicate that algorithm 6.7.3 is useful for graphs of common size.

6.8 Isomorphism of planar graphs

In the preceding section we have seen a quick method for finding a planar drawing of a graph, or proving its non-existence. Equally fast, i.e. in running time $O(n)$ where n is the number of vertices, is the possibility of determining whether two planar graphs are isomorphic. Let us remark here that, in contrast, no fast and reliable algorithm has yet been found for establishing the isomorphism of general graphs — see § 9.4 below.

We first present the following definition.

Definition 6.8.1. Let G_1 and G_2 be two planar graphs with planar drawings r_i and the corresponding cyclic orderings q_i of edges for $i = 1, 2$. We say that an isomorphism $f: G_1 \to G_2$ is compatible with the planar drawings r_1 and r_2 if for each directed edge h of the graph G we have $f(q_1(h)) = q_2(f(h))$. ●

The basis of the procedures we present below is a search for an isomorphism of topological planar graphs compatible with their planar drawings. Since such an isomorphism must preserve the geometrical relations of edges and vertices, which can be seen from the given drawings in the plane, the search is not too complicated.

As we observed in Chapter 1, a 3-connected planar graph can have only two different cyclic orderings of edges determined by planar drawings, and hence, each isomorphism of 3-connected planar graphs is compatible either with the given drawings, or with the given drawing of the first graph and the mirror image of the drawing of the second one. Consequently, a double application of the algorithm for finding an isomorphism of two 3-connected topological graphs compatible with their drawings will answer the question about the existence of a general isomorphism.

If our graphs are not 3-connected, they can have a substantially bigger number of planar drawings than two. Then it is appropriate first to decompose our graphs into their 3-connected components (by the algorithms sketched in § 6.5), then group these components into classes of mutually isomorphic graphs, and finally conclude the whole procedure by using the tree structure to describe the interrelationship of the components, as explained in § 6.6.

We now turn to the problem of isomorphism of topological planar graphs compatible with their drawings.

Definition 6.8.2. Let G be a topological planar graph, and let h be an orientation of an edge of G. Denote by $L(G, h)$ the ordered quadruple of numbers consisting of the degrees of the initial and terminal vertex of h, and the degrees of the faces lying to the right and to the left of h, respectively.

Given another topological planar graph G' and an orientation h' of an edge of G', then we say that h and h' are similar if $L(G, h) = L(G', h')$. ●

An isomorphism compatible with the planar drawings must preserve the quadruples $L(G, h)$, and therefore, the corresponding edges under the isomorphism must be similar. On the other hand, it is not difficult to find pairs of similar edges which do not correspond under any such isomorphism. For this reason we will introduce stronger concepts as follows.

Definition 6.8.3. Let G be a topological planar graph, and let q be a cyclic orientation of its edges. A directed edge (v, w) is said to be the *left continuation* of the directed edge (u, v) provided that $\{v, w\} = q(v, \{v, u\})$, in other words, provided that $\{v, w\}$ is the first edge after $\{u, v\}$ if we turn clockwise around the vertex v. A directed edge (v, w) is said to be the *right continuation* of (u, v) provided that $\{v, u\} = q(v, \{v, w\})$.

By an *admissible path* in the graph G is meant a directed path formed by directed edges $g_0, g_1, ..., g_k$ such that g_i is either a left or a right continuation of g_{i-1} for all $i = 1, ..., k$. Given another topological planar graph H and an admissible path $h_0, ..., h_k$ in H, we say that the above two paths are of the same type if for each $i = 1, ..., k$, the edge g_i is a left continuation of g_{i-1}, if, and only if, the edge h_i is a left continuation of h_{i-1} (which then holds analogously for right continuations). ●

An admissible path thus describes a motion in the topological planar graph which always continues either left-ways or right-ways. The way in which the left and right turns succeed each other determines the type of the path.

If there exists an isomorphism mapping the directed edge g to the directed edge h, then the isomorphism must necessarily map admissible paths of a given type starting at g and h onto each other. Consequently, the corresponding directed edges of the paths are similar. This motivates the following definition.

Definition 6.8.4. Let G and H be topological planar graphs, and let g and h be directed edges of G and H, respectively. The edges g and h are said to be indistinguishable if for two arbitrary admissible paths $g = g_0, g_1, ..., g_k$ in G and $h = h_0, h_1, ..., h_k$ in H of the same type, the edges g_k and h_k are similar. ●

The following theorem plays a key role for the problem of isomorphism compatible with planar drawings.

Theorem 6.8.5. *Let **G** and **H** be topological planar graphs, and let g and h be directed edges of **G** and **H**, respectively. Then the edges g and h are indistinguishable if, and only if, there exists an isomorphism of the graphs **G** and **H** compatible with their planar drawings and mapping g to h.*

The proof of this theorem will only be sketched here because, although its idea is rather graphic, the correct exposition is quite complicated. First, it is clear that if there exists an isomorphism compatible with the planar drawings and mapping g to h, then the edges g and h are indistinguishable.

Assume, conversely, that g and h are indistinguishable edges. We first show that for arbitrary admissible paths $g = g_0, g_1, ..., g_k$ in **G** and $h = h_0, h_1, ..., h_k$ in **H** of the same type, the edges g_k and h_k are also indistinguishable. In fact, given admissible paths $g_k, g_{k+1}, ..., g_{k+m}$ in **G** and $h_k, h_{k+1}, ..., h_{k+m}$ in **H** of the same type, then the paths $g_0, g_1, ..., g_{k+m}$ and $h_0, h_1, ..., h_{k+m}$ are also of the same type, and therefore, the edges g_{k+m} and h_{k+m} are similar.

The isomorphism to be presented is constructed as follows. First, assign to each other the corresponding edges and vertices of the circumference of the face lying to the left of g in **G**, and to the left of h in **H**, respectively. This is possible because the edges g and h are similar and hence the above faces have the same degree. The rotation along the face considered is a continuation of g or h which is always left-ways, and hence we obtain paths of the same type. It follows that the corresponding edges are always indistinguishable which, *inter alia*, guarantees that the corresponding vertices have the same degrees.

The above procedure is repeated, only this time for the faces lying to the right of the edges of the circumferences of the original faces. A rotation along the circumference is now understood as the admissible path with the right-ways continuations. The isomorphism is then always extended to the paths which are once left-ways and once right-ways continuations of the directed edges to which we have already assigned a value. Since we always assign to each edge of **G** an indistinguishable edge of **H**, no conflict can arise in the course of the construction.

The reader who does not believe that the above procedure is always correct can consult the paper [156]. ●

The above theorem is the basis for the following algorithm described by J E Hopcroft and R E Tarjan [156].

Algorithm 6.8.6. Isomorphism of topological planar graphs
Input data: connected topological planar graphs **G** and **H**.

Task: find an isomorphism of G and H compatible with their planar drawings.

Auxiliary variables:

$A_1, A_2, ..., A_p$, pairwise disjoint sets of directed edges of the graph G,

$B_1, B_2, ..., B_q$, pairwise disjoint sets of directed edges of the graph H (where $p \leq n$ and $q \leq n$, for the number n of vertices),

M, a set of pairs (i, D) where $1 \leq i \leq n$, the elements of which are called *types*,

X and Y, sets containing directed edges of G and H, respectively.

If, for example, $(i, L) \in M$, then the set of all left successors of the edges lying in A_i can cause a distinction of vertices lying in some of the sets A_j, and thus cause their separation.

1. [*Verify numerical parameters*] If the graphs G and H do not have the same number of vertices or edges, then they are not isomorphic, and the computation is terminated.

2. [*Distribute edges according to similarity*] For each directed edge g of the graph G determine the quadruple $L(G, g)$, and then distribute the edges into classes $A_1, ..., A_p$ in such a way that the following hold:

(i) g_1 and g_2 lie in the same class if, and only if, $L(G, g_1) = L(G, g_2)$, and

(ii) if $g_1 \in A_i$ and $g_2 \in A_j$ with $i < j$, then $L(G, g_1)$ precedes $L(G, g_2)$ in the lexicographic ordering.

The distribution is performed by first ordering the edges lexicographically (applying algorithm 5.4.4) after which it is easy to determine the separating boundaries of the individual classes.

Analogously, distribute the directed edges of the graph H into classes $B_1, ..., B_q$. Assuming that either $p \neq q$, or there is an index $i = 1, ..., p$ such that A_i and B_i have different numbers of elements, or have edges with different values of the functions $L(G, -)$ and $L(H, -)$, then the computation is terminated since G and H are not isomorphic.

3. [*Create sets of types*] Put M equal to the set of all pairs (i, L) and (i, R) for $i = 1, ..., p$.

4. [*Refine the distribution*] While $M \neq \emptyset$, perform the following operations 4a to 4c.

4a. [*Choose a type*] Choose a type $(i, D) \in M$, where $D = L$ or $D = R$, and delete (i, D) from the set M. Put $X := \emptyset$ and $Y := \emptyset$.

4b. [*Create division sets*] For each edge $g \in A_i$ insert into X the left continuation of the directed edge g in cases where $D = L$, and the right continuation in cases where $D = R$.

Analogously, create the set Y from the left or right continuations of the edges in B_i.

4c. [*Split the classes*] For each $g \in X$ perform the following operations:
(i) find j for which $g \in A_j$;
(ii) if the sets $A_j \cap X$ and $B_j \cap Y$ have different numbers of elements, terminate the computation: the graphs G and H are not isomorphic;
(iii) if $A_j \cap X \neq A_j$, do the following:
put $p := p + 1$; $A_p := A_j - X$; $A_j := A_j \cap X$; $B_p := B_j - Y$; $B_j := B_j \cap Y$;
if $(j, D) \in M$ insert (p, D) into M;
if $(j, D) \notin M$ do the following:
insert into M either the pair (j, D), in cases when A_j does not have more elements than A_p, or the pair (p, D) if A_j has more elements than A_p.
5. [*Determine the isomorphism*] If the computation has reached this stage, the corresponding pairs of sets A_i and B_i contain mutually indistinguishable edges, and the graphs G and H are isomorphic under an isomorphism compatible with their drawings. The isomorphism can be constructed by choosing a pair of indistinguishable directed edges g and h and then performing a simultaneous breadth-first search of both graphs with the initial edges g and h, by means of algorithm 4.2.6. ●

The proof of correctness of the above algorithm follows from the definition of indistinguishable edges and theorem 6.8.5. It is sufficient to observe that in step 4c(iii) if $(j, D) \in M$ it is not necessary to insert into M both of the pairs (j, D) and (p, D) to repeat the processing, but only either one of them. The choice made is motivated by the endeavour to reach a higher speed of computation, which we now estimate.

Theorem 6.8.7. *Algorithm* 6.8.6 *will process graphs of n vertices in running time* $O(n \log n)$.

Proof. A planar graph has less than $3n$ edges, and hence, less than $6n$ directed edges. In order to establish the degrees of vertices and faces, it is sufficient to search all edges, each of which is incident with two vertices and two faces. By theorem 5.4.5 the execution of steps 1, 2 and 3 thus requires running time $O(n)$.

One execution of steps 4a to 4c requires, assuming a good implementation, running time proportionate to the number of elements of the set A_i. Let g be a directed edge which lies in the set A_i at the instant at which the type (i, L) is being deleted from M, and then it appears in a certain set A_j at the instant at which (j, L) is being deleted from M. Then it follows from the description of step 4c(iii) that the number of elements of A_i at the first mentioned instant is at least twice the number of elements of A_j at the latter. It follows that the number of steps of computation such that g lies in a certain set A_k during the deletion of the type (k, L) from M is at most $\log(6n)$. The same is true for the pairs (i, R). Each edge g thus adds to the running time of step 4 an amount proportionate to $2\log(6n)$, and hence, step 4 takes running time $O(6n \cdot 2\log(6n)) = O(n \log n)$. ●

Algorithm 6.8.6 can be used to find general isomorphisms of 3-connected planar graphs if we make use of the fact that such graphs have only two planar drawings which are mutual mirror images.

Algorithm 6.8.8. Isomorphism of 3-connected planar graphs
Input data: 3-connected planar graphs G and H.
Task: find an isomorphism of G and H.

1. [*Planar drawing*] Using some of the algorithms of §6.7, determine planar drawings r of the graph G and s of the graph H.

2. [*Find an isomorphism*] Applying algorithm 6.8.5, try to find an isomorphism of the graphs G and H compatible with the given planar drawings r and s. If it does not exist, apply the same algorithm to try finding an isomorphism of the graphs G and H compatible with the drawing r and the mirror image of the drawing s. If this isomorphism also does not exist, then the graphs G and H are not isomorphic. ●

Algorithm 6.8.6 can be simply generalised to the searching for an isomorphism of both vertex-labelled and edge-labelled topological planar graphs. It is sufficient to extend, in definition 6.8.2, the quadruple $L(G, h)$ by three further numbers: the values of the edge h and of the initial and terminal vertex of h. The rest of the procedure is the same except that sometimes it is necessary to replace the sorting by distribution in step 2 by another, more general method. This, however, does not change the running time estimate of $O(n \log n)$.

Algorithm 6.8.6 can also be used to decide the existence of isomorphisms of a larger number of topological planar graphs. Given such graphs $G_1, G_2, ..., G_k$ then each of them is processed in the same manner as the graphs G and H in algorithm 6.8.6 as formulated above. The computation is performed simultaneously for all the given graphs, but it is not terminated when discrepancies of numerical characteristics are found in steps 1, 2 or 4c(ii). Instead, we distribute the given graphs into groups with the same values of the corresponding parameters, and each of these groups is processed until either it contains a single graph (which, then, is not isomorphic to any other of the given graphs), or we prove, by reaching step 4, that all the graphs in the group are pairwise isomorphic. The whole computation takes running time $O(N \log N)$ where N is the sum of the numbers of vertices of all graphs under study.

Soon after algorithm 6.8.6 was discovered, the same problem was solved by a quick algorithm [157] which is optimal up to a multiplication constant because it solves the task in running time proportionate to the number of vertices. The idea is quite simple. A certain collection of rules is given by which planar graphs can be reduced. Searching an isomorphism compatible with the planar drawing for two or more planar graphs, we reduce the graphs simultaneously using the given rules, and when the computation does not produce a discrepancy which would prove the non-existence of isomorphism, the result of all the reductions is a collection of regular polyhedra. The vertices and edges of those polyhedra are labelled in a specified way because we create lists for vertices and edges into which we encode the history of all reductions. Those lists are modified in the course of computation with the aim of obtaining a higher speed, and the modification is similar to that used in algorithm 6.6.1 where the lists $L(v)$ were used to modify the numbers $C(v)$.

Finally, we verify, for example by applying algorithm 6.8.6, whether the resulting regular polyhedra are isomorphic under an isomorphism compatible with the planar drawings and preserving the labelling of vertices and edges. As remarked in Chapter 1, it is possible to construct only five types of regular polyhedra (except the cycles for which the problem is easy), the largest of which, the icosahedron, has twenty vertices. Therefore, the length of the last stage of computation does not at all depend on the sizes of the original graphs, only on their number, which is usually two.

We will discuss the whole algorithm only informally. In the course of the reductions, multiple edges, loops and vertices of degree 1 can be created. They are immediately removed, but information concerning this is stored in the lists describing the history of the computation. Priorities are assigned to the further possible reductions, and one always executes the one that has the highest priority. As long as there exist two vertices of different degrees, or two faces of different degrees, at least one of the reduction rules is actually applicable, and hence, the final result is a regular polyhedron.

The reductions are based on the existence of at least one vertex of degree equal to or smaller than 5. Moreover, if the graph is reduced in such a way that all vertices have the same degree, then the graph is either a cycle, or it contains a face of degree at most 5; see Chapter 1. The reduction of the highest priority removes a vertex of degree d, where $d \leq 5$, which has no neighbour vertex of degree d. This is performed by 'splitting' the vertex into a polygon which is spanned in between the neighbours of the vertex to be deleted in the manner indicated in figure 27. Eventual multiple edges are removed. The vertices with the lowest degrees have the highest priority for the reduction under consideration. It is important that when

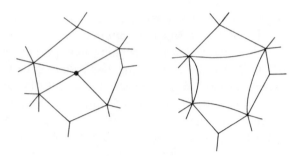

Figure 27.

deleting the vertices of degree d, the 'working regions' are mutually separated by vertices of degrees different from d, and hence, the individual reductions do not influence each other and they can be performed in an arbitrary order for a chosen value of d.

The list of further reductions is rather extensive, and it contains a number of exceptions. We thus do not present it here. It seems that the whole procedure is, due to its complexity, not as suitable for practical implementations as algorithm 6.8.6. Although it works in running time proportionate to the number of vertices, the constant of proportion is rather large, and hence the influence of the slowly increasing factor of $\log n$ in theorem 6.8.7 will be outweighed in practical applications in cases of smaller graphs. Practical experiments in this direction have not been made, however.

Finally, we will exhibit a way of searching the isomorphism of planar graphs of smaller degree of connectedness than 3. We first consider the case of 2-connected graphs. We proceed by first decomposing the graph into its 3-connected components using, for example, the procedure mentioned in §§ 6.5 and 6.7. We then create a graph the vertices of which are the components thus obtained and two of these vertices are connected by an edge if, and only if, the corresponding 3-connected components have a common vertex. It is not difficult to prove that the resulting graph is a tree. The tree of 3-connected components is then processed on two levels. Firstly, the 3-connected components are distributed to groups of pairwise isomorphic components and they are then labelled in such a way that two components obtain the same value if, and only if, they are isomorphic. This can be done, for example, by algorithm 6.8.8 except that those components which are represented by cycles are processed separately. Secondly, the trees of components are processed by algorithm 6.6.4 with respect to the labelling of the vertices by numbers labelling the isomorphism class of the component represented by the vertex. Applying algorithm 6.6.4

and determining isomorphism of the 3-connected components by the algorithm of J Hopcroft and J K Wong mentioned above, using reductions, it is possible to determine isomorphism of 2-connected planar graphs in running time proportionate to the number of vertices. An application of algorithm 6.8.8 then yields the running time estimate $O(n \log n)$.

Searching for isomorphism of connected planar graphs is performed in a similar way: the graph is decomposed into a tree of 2-connected components which are then distributed into groups in the manner described in the preceding paragraph, and the evaluated tree of components is processed by algorithm 6.4.4. The difference lies in the simpler manner of decomposition of components and a simpler description of their interrelation, since two distinct 2-connected components can meet in a single vertex only, in contrast to the 3-connected ones. Finally, the search for isomorphism of general graphs is performed by decomposing the graphs into connected components, distributing the components according to isomorphism, and then concluding that the graphs are isomorphic if, and only if, they have the same number of components in each of the groups.

Globally, we see that the existence of isomorphism of two planar graphs can be established in running time proportionate to the number of vertices by applying (1) methods for searching components of various degrees of connectedness, (2) algorithms creating planar drawings, (3) methods for determining isomorphism of trees and of planar graphs, and (4) the theorem on the uniqueness of the planar drawing of a 3-connected graph. However, the resulting algorithm is so complicated that it will hardly ever be practically used. On the other hand, searching for it and finding it has made a great contribution to the algorithmic theory of graphs, and besides, the individual tricks used in the algorithm have often been found to be useful.

By generalising the algorithms of §§ 6.7 and 6.8 we can obtain algorithms which determine in polynomially bounded running time whether a graph can be drawn onto a surface of a higher order, e.g. onto a torus, or into the projective plane, and which also determine isomorphism of general graphs drawn on those surfaces, see [101, 102, 188 and 210]. In all these cases the degree of the polynomial bounding the length of the computation is linearly proportionate to the degree of the surface.

Exercises

1. Construct an edge-labelled graph **G** such that the depth-first search will not determine the shortest path from the initial vertex to some other vertex of the graph.

2. Show that Dijkstra's algorithm can lead to a wrong result in a generally labelled acyclic directed graph.
3. Design a procedure which in a generally labelled graph would either find a cycle with a negative sum of edge, or prove its non-existence.
4. In the set of all non-negative numbers with the element $+\infty$ added we define operations $\oplus$ and $\otimes$ as follows:

$$a \oplus b = \min (a, b) \quad \text{and} \quad a \otimes b = a + b.$$

Further, denote the element $+\infty$ by $\textcircled{0}$ and the element 0 by $\textcircled{1}$. We introduce 'addition' and 'multiplication' of n by n matrices (a_{ij}) and (b_{ij}) of elements of the considered set by the usual rule:

$$(c_{ij}) = (a_{ij}) \oplus_n (b_{ij}) \quad \text{if} \quad c_{ij} = a_{ij} \oplus b_{ij}$$

and

$$(d_{ij}) = (a_{ij}) \otimes_n (b_{ij}) \quad \text{if} \quad d_{ij} = (a_{i1} \otimes b_{1j}) \otimes \ldots \otimes (a_{in} \otimes b_{nj}).$$

Furthermore, we define the 'unit matrix' $\mathbf{I}_n = (e_{ij})$ by the following rule:

$$e_{ij} = \textcircled{0} \quad \text{if} \quad i \neq j, \; e_{ij} = \textcircled{1}.$$

Let G be a directed graph in which the edges are labelled by integers. If $v_1, \ldots, v_n$ are the vertices of G, then we put

$$a_{ij} = \begin{cases} \text{the labelling of the edge } (v_i, v_j) \text{ is an edge,} \\ +\infty \text{ otherwise.} \end{cases}$$

Show how, given a matrix $\mathbf{A} = (a_{ij})$, one can compute the following matrix:

$$\mathbf{B} = \mathbf{I}_n \oplus_n \mathbf{A} \oplus_n (\mathbf{A} \otimes_n \mathbf{A}) \oplus_n (\mathbf{A} \otimes_n \mathbf{A} \otimes_n \mathbf{A}) \oplus_n \ldots$$

in running time $O(n^3)$, and prove that the resulting matrix is the matrix of distances of the vertices of the graph G.

5. In the set $\{0, 1\}$ define the following operations:

$$a \oplus b = \min (a + b, 1) \quad \text{and} \quad a \otimes b = ab$$

and put $\textcircled{1} = 1$ and $\textcircled{0} = 0$.

Let $\mathbf{A}$ be the incidence matrix of a graph G. Show that a coefficient b_{ij} of the matrix $\mathbf{B}$ defined as in the final part of exercise 4 is 1 if there is a directed path from the ith to the jth vertex of the graph G.

Suppose an algorithm is given making it possible to perform the operation $\otimes_n$ defined by means of the operations $a \oplus b = \min (a + b, 1)$ and $a \otimes b = a \cdot b$ in running time $O(f(n))$. (An algorithm is known with $f(n) = n^{2.49}$, a simple procedure with $f(n) = n^{2.81}$ is given in [1] or [200], and there is a practically implemented algorithm with $f(n) = n^3/\log n$, see [45].) Present a determination of the matrix **B**, the so-called transitive closure of the graph G, in running time $O(f(n))$.

6. Let the directed edges of a directed graph G represent activities necessary to perform a certain project, and let their evaluation express the time required for the corresponding activities. The terminal vertex of an edge g is equal to the initial vertex of an edge h if, and only if, the activity corresponding to h cannot begin before the termination of the activity which corresponds to g. Further, let G contain vertices i and t representing the initialisation and the termination of the whole project with the property that every directed edge of the graph G lies on a directed path from i to t.

Verify that the longest path from i to t in the graph G determines the shortest time in which the project can be realised. What would the existence of a cycle in the graph G mean under the present interpretation?

The above application of the search of the longest path is the basis of the methods CPM and PERT which have successful applications in operational analysis.

7. Design an algorithm which in an edge-labelled graph finds a closed sequence passing through all vertices of the graph and whose labelling is not worse than the double of the corresponding value of the minimum closed sequence. Verify that running time $O(n \log n)$ is sufficient for this task.

8. Verify that in a non-negatively labelled graph there exists at least one subgraph fulfilling the conditions (a) and (b) of lemma 6.2.1, which is a tree. Are there any restrictions on the form a graph fulfilling the conditions (a) and (b) of lemma 6.2.1 which is a subgraph of a generally labelled graph?

9. How can one find, in a generally labelled graph, a subgraph fulfilling the conditions (a) and (b) of lemma 6.2.1, and how long would it take?

10. How can one find, in a generally labelled graph, a subtree containing all vertices and having the minimum sum of the labellings of edges?

11. Suppose that to each directed edge h of a network S a real number $w(h)$ is assigned, called the price of the edge. By the price of a flow f we understand the sum of the numbers $f(h) w(h)$ over all edges of the network. For each directed edge $h = (x, y)$ of the graph $R(f)$ put $h^r = (y, x)$, and denote by $w(f, h)$ either the number $f(h) w(h)$ when h is an edge of the network S but h^r is not an edge of S, or the number $-f(h^r) w(h^r)$ when h^r is an edge of S but h is not, or the number $f(h) w(h) - f(h^r) w(h^r)$ when both h and h^r are edges of the networks.

Prove that the price of a flow can be decreased while keeping the magnitude of the flow if, and only if, there exists a cycle in the graph $R(f)$ whose vertices $h_1, h_2, ..., h_k$ fulfil

$$w(f, h_1) + w(f, h_2) + ... + w(f, h_k) < 0.$$

12. Using the above exercises 3 and 11, design an algorithm for finding the maximum flow which has the minimum price.

13. Describe an algorithm for finding the minimum cut in a network.

14. Using exercise 12, find, in an edge-labelled bipartite graph, a maximum matching with the largest possible sum of the labelling of edges.

15. Let G be a graph. Define a network on the vertices (x, i) where x is a vertex of the graph G and $i = 1, 2$, whose edges have either the form $((x, 1), (x, 2))$ for any vertex x of the graph G, or the form $((x, 2), (y, 1))$ where $\{x, y\}$ is an edge of the graph G. All capacities of edges are equal to 1. Prove that, for arbitrary vertices v and w of the graph G, the above network with the source $(v, 1)$ and the target $(w, 2)$ has the maximum flow equal to the minimum number of vertices which have to be deleted from G in order to separate the vertices v and w into distinct components of the new graph.

16. Let f and g be self-maps of the set X. How can one find a bijective self-map φ of the set X such that for each $x \in X$ we have $\varphi(f(x)) = g(\varphi(x))$? Is it possible either to find φ or prove its non-existence in running time proportionate to the number of elements of the set X?

7

NP-complete problems

There are a number of problems in graph theory and operational analysis for which we are not able to guarantee that an optimum solution will be found even when using the most efficient computers. Most of those problems belong to the class of so-called *NP*-complete problems. There is a widely spread (although not yet verified) belief that for the solution of such problems no algorithms exist whose speed would be comparable to that of the algorithms of the preceding chapter. Some authors even call *NP*-complete problems unsolvable from the practical point of view.

The aim of the present chapter is first of all to explain the reasons for this pessimistic attitude towards the question of the computational complexity of *NP*-complete problems, and then to show some important problems which belong to that class, e.g. colouring of graphs, searching a hamiltonian circuit, problems of transport and planning based on the above problems, and a number of further problems.

7.1 Introduction

In the last decade, much success has been achieved in the realm of investigation of the interrelationship of the computational complexity of a lot of difficult problems. As an illustration we will study the relationship between finding an independent set of a graph and verifying logical formulae.

Example 7.1.1. Let a graph G and a natural number k be given. We will construct a formula in the conjunctive normal form $f = f(u_1, ..., u_n)$, such that f is satisfied if, and only if, the set $\{x_i | u_i = \text{true}\}$ is an independent set of the graph G of at least k elements.

It is sufficient to take as f the conjunction of formulae of the form $\neg u_i \lor \neg u_j$ where $\{x_i, x_j\}$ is an edge of the graph G, and the formula $g(u_1, ..., u_n)$ which is satisfied if, and only if, at least k of its variables have the value true. In fact, if the expression $\neg u_i \lor \neg u_j$ is satisfied, then at least one of the variables u_i and u_j must be equal to false, and hence, at most one of the end-vertices of the edge $\{x_i, x_j\}$ can be in the set $\{x_i/u_i = \text{true}\}$.

One of the simplest, although not the best by far, methods of construction of the formula g is the following: use auxiliary variables w_{pq} for $p = 1, 2, ..., n$ and $q = 0, 1, ..., n$, and let g be the conjunction of the following formulae:

(1) $\neg w_{pq} \lor \neg w_{pr}$ for $p = 1, 2, ..., n$ and $0 \leq q < r \leq n$;

(2) $(\neg u_1 \lor w_{11}) \land (u_1 \lor w_{10})$;

(3) $\neg u_p \lor \neg w_{p-1,q-1} \lor w_{pq}$ for $p = 2, 3, ..., n$ and $q = 1, 2, ..., n$;

(4) $u_p \lor \neg w_{p-1,q} \lor w_{pq}$ for $p = 2, 3, ..., n$ and $q = 0, 1, ..., n$;

(5) $w_{nk} \lor w_{n,k+1} \lor ... \lor w_{nn}$.

The formulae for the first line (1) above are satisfied if, and only if, for each p there exists at most one q with $w_{pq} = \text{true}$. The formulae of lines (2), (3) and (4) are satisfied if, and only if, w_{pq} is equal to true, i.e. if and only if, exactly q of the variables $u_1, ..., u_p$ have the value true: if u_1 is true or false, then we must have $w_{11} = \text{true}$ or $w_{10} = \text{true}$, respectively (due to line (2)), and if $u_p = \text{true}$ and $w_{p-1,q-1} = \text{true}$, then necessarily $w_{pq} = \text{true}$ (due to line (3)), and if $u_1 = \text{false}$ and $w_{p-1,q} = \text{true}$, then $w_{pq} = \text{true}$ (due to line (4)). Thus, if q of the variables are equal to true, then $w_{nq} = \text{true}$, and line (5) forces the inequality $q \geq k$. ●

The search for an independent set of size at least k can be performed, by means of the above example, as follows: construct the formula q and apply an algorithm investigating the satisfiability of logical formulae. When trying to find quick algorithms, it is thus desirable to restrict our attention to the problem of satisfiability, since by solving it we would get free of cost a comparatively quick algorithm for searching for large independent sets. A pessimist, on the other hand, would try to prove the non-existence of a quick algorithm for searching for independent sets of a prescribed size in order to show, when he succeeds, that the investigation of satisfiability cannot be performed quickly.

It turns out, moreover, that the relationship of the two problems is two-sided, as the following example shows.

Example 7.1.2. Let a logical formula f be given in its conjunctive form,

$$f(u_1, u_2, ..., u_n) = \bigwedge_{i=1}^{m} \bigvee_{j=1}^{m_i} a_{ij}$$

where each a_{ij} has the form u_k or $\neg u_k$ for a suitable k. We will construct a graph G in which an independent set of size m exists if, and only if, the formula f is satisfiable. The vertices of the graph G are elements x_{ij} for $i = 1, 2, ..., m$ and $j = 1, 2, ..., m_i$, and two vertices x_{ij} and x_{pq} are connected by an edge if either

$$i = p \quad \text{and} \quad j \neq q,$$

or if there exists k such that

$$\text{either} \quad a_{ij} = u_k \quad \text{and} \quad a_{pq} = \neg u_k$$

$$\text{or} \quad a_{ij} = \neg u_k \quad \text{and} \quad a_{pq} = u_k.$$

If $f(u_1, ..., u_n) = $ true, then for each i we can choose a $j = j(i)$ such that $a_{ij} = $ true. It follows easily that the set $x_{i, j(i)}$, $i = 1, 2, ..., m$, is independent in the graph G, and it has m elements. Conversely, let M be an independent set of the graph G. When x_{ij} and x_{pq} both lie in M and $i = p$, we have, necessarily, $j = q$, and hence, $x_{ij} = x_{pq}$. Thus, if M has n elements, then for each i there must exist exactly one $j = j(i)$ such that $x_{i, j(i)} \in M$.

Now, if $a_{i, j(i)} = u_k$, put $u_k = $ true, and if $a_{ij(i)} = \neg u_k$, put $u_k = $ false. In both cases, $a_{i, j(i)} = $ true. It follows from the independence of the set M and the definition of the graph G that it cannot happen that the value of any variable is defined in two distinct ways. Defining arbitrarily the values of the variables u_i which have not yet been determined, we obtain $f(u_1, u_2, ..., u_n) = $ true. ●

Thus, a quick algorithm for searching for independent sets of a prescribed size would also make it possible to decide quickly about the satisfiability of logical formulae in conjunctive normal form.

It follows from examples 7.1.1 and 7.1.2 that the computational complexity of the two problems under consideration is not very different, and when we use a two-valued classification (the existence or non-existence of a quick algorithm) then they will both be classified in the same way.

In 1972 R M Karp [168], presenting the results of his own and other authors' research, listed a further nineteen problems which have, in the above-mentioned sense, the same computational complexity as the search for an independent set, or as the decision of satisfiability of formulae. That list contained important problems of graph theory (e.g. the existence of a colouring by k colours, the existence of a hamiltonian circuit or path) and of operational analysis (e.g. problems of integer programming, the travelling salesman problem, and others). At the present time, several hundreds of such problems are known, see [17]. These problems are said to be *NP*-complete, for reasons to be seen in § 7.4.

To use a graphic metaphor, in the 1960s it still seemed that the above problems resembled heavy boulders which we would have liked to move and for which there was hope that at least some of them would show a weak spot making such a move possible. By the beginning of the 1970s, however, it had turned out that those boulders are interconnected in such a way that we are actually attempting to move a whole mountain. It is thus hardly surprising that no one has yet managed to move it in either direction, i.e. neither do we know a quick and precise algorithm for any of those problems (which would mean finding quick algorithms for hundreds of difficult problems at once), nor do we know a proof that such algorithms do not exist (which would show that none of the NP-complete problems can be coped with). Today, it is firmly believed that sufficiently quick algorithms for solving NP-complete problems will never be found.

This does not mean, however, that we are completely impotent to deal with such problems, but when designing methods of solution of those problems, we are forced to set lesser goals than that of finding an algorithm which would always quickly find the optimum solution of the problem in which we are interested. Such a lessening of goals is primarily possible in the case of optimisation problems when we can be satisfied with an approximate solution. This has been successful first of all for problems of a numerical type which will be discussed in § 7.8; see also [124 and 158].

On the other hand, for optimisation problems of a purely combinatorial type, e.g. colouring of graphs using the smallest number of colours, the task of finding a good approximate solution is often as difficult as that of finding the optimum solution, see e.g. [125]. For a decision problem where the task is to decide whether the input data have a certain property, it is even difficult to describe what is meant by an approximate solution. In those cases we must further lower our goal and try to find algorithms which are satisfactory from the probabilistic point of view, i.e. the correct solution or its approximation will be found in a majority of cases of application of the algorithm, but we cannot eliminate the possibility of an erroneous or inexact answer, or of an overly prolonged computation.

The last approach yields good results even for very difficult combinatorial problems, which is verified by practical experience with the heuristic methods as described in Chapter 9, as well as by theoretical considerations as shown in Chapter 10.

Thus, the theory of NP-complete problems is, in spite of its negative character, of practical importance, since it makes it possible to evaluate practically the degree of difficulty of the problem we are trying to solve, and consequently to decide whether we are going to try and find an exact algorithm, an approximate solution or a heuristic method which we believe will be successful in the case to which we apply it.

7.2 Problems solvable in polynomial-bounded time

In § 7.1 we have often used the expression 'quick algorithm' without specifying what we mean. The reader certainly has an intuitive idea of which algorithms are quick and which are slow but these ideas are usually rather individual, and too imprecise to enable an exact investigation of the speed of algorithms.

For this reason, we will attempt to define quick algorithms precisely, although we are fully aware of the fact that the boundary between quick and slow is so fuzzy that a number of well-founded objections can be raised against any formal definition. In theoretical considerations, the most common approach is that based on the following observation first formulated in the paper [87].

The computation time required to process input data of size n by algorithms considered to be quick, e.g. the procedures exhibited in Chapter 6, usually have an upper bound by a function of the form n, $n \log n$, n^2, n^3, etc. Thus, in each case, a polynomial function in the variable n can be used as an upper bound.

On the other hand, the typical representative of slow algorithms, the method of 'brute force', requires at least 2^n steps if all subsets of the given n-point set of inputs are considered, or $n!$ steps if all permutations are considered, or n^n steps if all self-maps are considered. Thus, here the upper bound of the number of steps is at least exponential, and this remains true even when we use the optimisation method described in Chapter 4, § 4.3.

We will thus consider an algorithm to be quick if there exists a polynomial function p such that the number of steps required for the computation is bounded by the number $p(n)$ where n is the size of input data. If such a polynomial does not exist, the algorithm will be considered to be slow.

To illustrate the difference between algorithms working in polynomially bounded time and exponential algorithms, we will show, in the subsequent table, times needed to compute input data of size n in the case where the number of operations which are to be performed is given by the function $f(n)$ and the performance of one operation takes one microsecond.

An even better insight into the difference between polynomial and exponential algorithms can be obtained from the second table which shows the expansion of the range of computable problems corresponding to using a computer of speed 10 times, 100 times and 1000 times, respectively, higher than the original computer, assuming that originally it was possible to compute input data of size $n = 1000$ within the given time limit.

The following tables demonstrate that a typical feature of exponential algorithms is the existence of an upper bound on the size of input data,

Number of operations	$n = 20$	$n = 40$	$n = 60$	$n = 80$
n	20 μs	40 μs	60 μs	80 μs
$n \log n$	86 μs	0.2 ms	0.35 ms	0.5 ms
n^2	0.4 ms	1.6 ms	3.6 ms	6.4 ms
n^3	8 ms	64 ms	0.22 ms	0.5 ms
n^4	0.16 s	2.56 s	13 s	41 s
2^n	1 s	11.7 days	36 600 years	3.6×10^9 years
$n!$	77 000 years			

Number of operations	$n = 100$	$n = 200$	$n = 500$	$n = 1\,000$
n	0.1 ms	0.2 ms	0.5 ms	1 ms
$n \log n$	0.7 ms	1.5 ms	4.5 ms	10 ms
n^2	10 ms	40 ms	0.25 ms	1 s
n^3	1 s	8 s	125 s	17 min
n^4	100 s	27 min	17 h	11.6 days
2^n				
$n!$				

$f(n)$	Speeding up			
	once	10 times	100 times	1 000 times
n	100	1 000	10 000	100 000
$n \log n$	100	702	5 362	43 150
n^2	100	216	1 000	3 162
n^3	100	215	464	1 000
n^4	100	177	316	562
2^n	100	103	106	109
$n!$	100	100	100	101

over which the problem is unsolvable even when the speed of the computer is increased by several orders.

It is possible to object that an algorithm requiring computation time n^{100} or $2^{100}n$ is also impractical, although the time bound is polynomial, and in the latter case even linear. In turns out, however, that if an 'everyday' problem has an algorithm with a polynomial time bound, then the polynomial has usually degree not exceeding 3 or 4, and also the coefficients are usually 'reasonable'. The objection is nevertheless valid, as illustrated by a recent algorithm for solving the problem of linear programming which has polynomial time bound, see [69]: that algorithm cannot compete, at least not in the present state of our knowledge, with the well-known simplex method, although for the latter the worst-case approximation is exponential; see [177 and 273]. We will thus use the expression 'quick algorithm' in intuitive considerations only, and whenever the precision of exposition is stressed, we shall always speak about algorithms working in polynomial-bounded time.

We conclude with a definition.

Definition 7.2.1. We denote by P the class of all languages J such that there exists a polynomial p for which the time complexity of the language J is at most p. ●

7.3 Problems solvable non-deterministically in polynomial-bounded time

Although there is a vast difference between the complexity of the problems investigated in Chapter 6 and that of the NP-complete problems, all those problems, as well as some others for which the classification is not yet known (e.g. the problem of searching for an isomorphism of graphs) have a property in common: once the solution has been found, it is easy to prove its correctness.

For example, if we are to colour a given graph with three colours, then it might be extremely difficult to find such a colouring, but once it has been found, then it is sufficient to label all vertices by their appropriate colours, and in order to check the correctness of the given solution, all we have to do is to take all edges and to check the colouring of their end-vertices. Similarly, if we write down the itinerary of a travelling salesman in a labelled graph, then the proof that the path is not longer than a prescribed limit is reduced to simple additions and searches in the matrix of distances.

It is interesting to observe the pronounced asymmetry between affirmative and negative solutions: if we have to prove to a mistrusting boss that the solution he requires, for example the design of a timetable with the upper bound of five years, does not exist, then our task is no easier than that of deciding whether or not a solution exists, in fact.

The simplicity of proving the correctness of solutions is a feature characteristic of decision problems and of constructive ones only, whereas in order to prove the optimality of a solution with value k we have to perform two steps: first we have to prove that our solution is admissible and has value k, and then we must show that there exists no admissible solution with value better than k. The first part is easily verified, but nothing like that can be said about the latter one, unless we can rely on a piece of information comparable with theorem 6.3.8 (which expresses the size of a maximum flow in a network as the minimum size of a cut).

To each optimisation problem of the form 'find an admissible solution of the smallest value' we can assign the decision problem 'determine whether there exists an admissible solution of value k', and from the complexity of solution of the latter we can judge the complexity of the original problem. The latter problem has the property of simple proof of correctness.

The complexity of a proof of correctness of an affirmative solution will play the role of delimitation of a class of problems the importance of which will be seen in the subsequent sections of this chapter. In order to define the class, we shall introduce the concept of a non-deterministic algorithm. As we shall see, the length of a non-deterministic 'computation' corresponds to the time required for a verification of the correctness of a solution. We will thus speak about problems solvable non-deterministically in polynomial-bounded time. We first have to explain

— what is a non-deterministic algorithm?
— how does a non-deterministic algorithm 'compute'?
— how is the result of such a 'computation' determined?
— how is the length of such a 'computation' specified?

A non-deterministic algorithm differs from a deterministic one by the feature that at some (or all) steps it is possible to decide quite freely between several possible continuations of the computation. The computation is thus not determined by the initial configuration, and instead of a computation we should rather speak about the system of all possible computations.

As an illustration, we first describe a simple algorithm for deciding whether there is an independent set X of at least k elements in a given graph G of vertices $v_1, v_2, ..., v_n$.

1. [*Initiation*] $X := \emptyset$.

2. [*Non-deterministic construction of* X] For $i := 1, 2, ..., n$ perform the following instruction: put either $X := X \cup \{v_i\}$, or $X := X$.

3. [*Verify independence of* X] If there exist vertices v_i and v_j in X such that $\{v_i, v_j\}$ is an edge, then *REJECT*.

4. [*Count the number of vertices*] If X has at least k elements, then *ACCEPT*, else *REJECT*.

This algorithm should reply *ACCEPT* or *REJECT* according to whether the required independent set exists or not. The algorithm is naturally worthless when we actually have to find an independent set of the required size because we would have to go through all, or a majority, of the possible performances of step 2, and since these possibilities correspond to all possible subsets of the vertex set of the graph *G*, their number is 2^n. We can also use the probabilistic interpretation of the above procedure which assumes that in step 2 we decide according to some stochastic rule and then we ask what is the probability of finding a correct solution, but again we get nothing really useful since the actual probability of correctness is usually very small. This, however, is no defect because we do not attempt to find a solution, but to create a concept which would make it possible to evaluate exactly the complexity of proving correctness of an affirmative solution.

If in the above example the solution being sought actually exists, then there also exists a computation which agrees with the non-deterministic scheme and returns the answer *ACCEPT*. If we manage to find the solution (which can be difficult), then we can use it at any time for an immediate proof of correctness of the solution.

The above example has clarified the concept of a non-deterministic algorithm. Let us now try and think about the performance of a 'computation' and determination of the 'result of computation'. A non-deterministic algorithm does not determine, for given input data, the unique succession of instructions, but rather a complex system of admissible computations which return different results. Not until we have seen the total collection of all computations can we decide whether the input has been accepted or rejected.

We will thus restrict ourselves to defining the concept of an admissible computation, but we shall define the global 'result of computation' which will be *ACCEPT*ed if, and only if, at least one admissible computation accepts the given input data. In order to be able to introduce the global result, we restrict our attention to problems which require the answers *ACCEPT* or *REJECT* as their solution.

For a formally precise definition we shall use acceptor, as defined in 2.3.2, enriched by an instruction *CHOOSE* which means an unconditional jump to one of two given steps in the program.

Definition 7.3.1. A *non-deterministic acceptor* differs from a (deterministic) acceptor only by admitting the instruction *CHOOSE L1, L2*, where *L1* and *L2* are labels. ●

Definition 7.3.2. An *admissible computation of a non-deterministic acceptor* M is an infinite sequence $C_0, C_1, \ldots$ of configurations of the acceptor M such that C_0 is an initial configuration of the acceptor M, and for each *i* the following holds: if the contents of the accumulator in the configuration C_{i-1} is k, and if the kth instruction of the program of the acceptor M is *CHOOSE* $L1, L2$, then the configurations C_{i-1} and C_i differ only in the contents of the program counter which in C_i is either $L1$ or $L2$ and if the kth instruction is not *CHOOSE*, then the relation of C_{i-1} and C_i fulfils the conditions described in §2.1. ●

The performance of the instruction *CHOOSE* $L1, L2$ corresponds to the performance of either *JUMP* $L1$, or *JUMP* $L2$, according to an individual choice.

Definition 7.3.3. We say that a word w is *accepted* by the non-deterministic acceptor M if there exists an admissible computation of the acceptor M whose initial configuration is determined by w and which returns the answer *ACCEPT*. If such a computation does not exist, we say that M *rejects* the word w.

We say that the acceptor M accepts the language **J** if for each word w we have $w \in$ **J** if, and only if, M accepts the word w. ●

Now we still have to describe the way in which the length of the 'computation' in the global sense will be determined. Imagine that all the admissible computations of the acceptor M determined by the word w are performed simultaneously. At the instant when one of the computations halts by performing the instruction *ACCEPT* we know that the word w is accepted, and we need not perform the remaining parts of the other computations (and, in fact, we can even admit that some of the other computations would never halt). However, if the word w is always rejected, the result will not be known until the instant at which all admissible computations have been done.

Definition 7.3.4. We say that a non-deterministic acceptor M processes a word w in running time at most t in one of the following cases:

either M accepts w and there exists at least one admissible computation of the acceptor M determined by the input word w which halts performing the instruction *ACCEPT* after at most t steps, or

the acceptor M rejects w and all admissible computations of the acceptor M determined by the input word w halt after at most t steps.

We say that a non-deterministic acceptor M works in polynomial-bounded time if there exists a polynomial p such that an arbitrary word of length n is processed by the acceptor M in running time at most $p(n)$.

We say that M processes a word w using memory space at most m if for a certain time t the following holds: the acceptor M will process the word w in running time t, and in the course of the performance of the first t steps of any of the admissible computations of M determined by the input word w no instruction is performed in which an instant address of an operand would be larger than M. ●

The following is a non-deterministic variant of definition 7.2.1.

Definition 7.3.5. We denote by NP the class of all languages **J** such that there exists a non-deterministic acceptor which works in polynomial-bounded time and accepts the language **J**. ●

The notation P and NP for the above classes is the abbreviation of the words 'polynomial' and 'non-deterministically polynomial'. Since a deterministic acceptor is a special case of a non-deterministic acceptor, we have the following

Theorem 7.3.6. $P \subset NP$. ●

The class P consists of problems which are considered to be practically solvable, and it contains (after an eventual reformulation in the form of decision problems) all problems mentioned in Chapter 6. On the other hand, a majority of problems which are of practical importance and which have to be solved often are members of the class NP. We have already mentioned the fact that it is not yet known whether $P = NP$, or $P \neq NP$. This question is often considered to be one of the most important open problems of modern mathematics, and it has remained unsolved in spite of an immense effort by a great number of mathematicians. It is generally believed that the two classes are not equal, since otherwise there would exist, for the hundreds of already known NP-complete problems, algorithms with polynomial-bounded running time. Some authors even believe that neither the equality nor the non-equality can be proved in the usual logical systems. The possible relationships of the classes P and NP are depicted in figure 33 in the next chapter. This figure also includes some other classes which we will study later.

7.4 *NP*-complete problems

From the historical point of view, the following result of S A Cook [75] had been of major importance: Cook showed that every problem which is a member of the class NP can be reduced in polynomial-bounded time to the problem of determination of satisfiability of logical formulae in conjunctive normal form, analogous to example 7.1.1 where we reduced the search of an independent set of a given size to that problem. The complexity of the problem of determination of satisfiability thus has the highest possible degree within the class NP. Due to the converse transformation exhibited in example 7.1.2 we see that the same maximum complexity is reached by the problem of searching an independent set.

Cook's theorem justifies the effort that we put into the introduction of the class NP in the last section: it is often obvious at first sight that a certain problem belongs to the class NP, as we observed at the beginning of § 7.3, but, nevertheless, an important corollary follows from this fact: the proof of existence of a reduction of the given problem to the problem of determination of satisfiability. Cook's result has initiated the search for further problems of the same complexity as that of the determination of satisfiability of formulae.

We shall first present an exact definition of the concept of reduction of one problem to another in polynomial-bounded time. Problems are again understood to be decision problems and are formalised as languages. Throughout the present section we assume the restriction introduced in 2.2.3.

Definition 7.4.1. We say that a language J_1 is reducible to a language J_2, and we denote this by $J_1 \lhd J_2$, if there exists a deterministic transducer M working in polynomial-bounded time and such that for every word w we have $w \in J_1$ if, and only if, $M(w) \in J_2$. ●

By means of this definition it is now possible to describe the concept of NP-completeness quite exactly.

Definition 7.4.2. A language J is said to be NP-complete provided that $J \in NP$ and for each language J' with $J' \in NP$ we have $J' \lhd J$. ●

Before proving the NP-completeness of the problem of determination of satisfiability of logical formulae, we will present some properties of the reduction of problems introduced above.

Theorem 7.4.3. *Reducibility in the sense of definition 7.4.2 is transitive, i.e.* $J_1 \lhd J_2$ *and* $J_2 \lhd J_3$ *imply* $J_1 \lhd J_3$.

Proof. Let M_1 be a transducer realising the reduction $J_1 \lhd J_2$, and M_2 that realising the reduction $J_2 \lhd J_3$; then a transducer realising the reduction $J_1 \lhd J_3$ can be obtained as follows: in the transducer M_1 each instance of the instruction *STOP* is substituted by the instruction of unconditional jump to the first instruction of the program of the transducer M_2 (concatenating the latter program to the program of M_1). Since in a random access machine program, as described in § 2.1, labels mean the absolute positions of the instructions in the program, it is also necessary to correct the labels in the second program which are shifted in the composed transducer from their original positions. If the transducers M_1 and M_2 work in time bounded by the polynomials p and q, respectively, then the composed transducer will work in time bounded by the polynomial $r(n) = q(p(n))$. ●

Theorem 7.4.4. *If* $J_1 \lhd J_2$ *and* $J_2 \in NP$, *then* $J_1 \in NP$.

Proof. The fact that $J_1 \in NP$ can be verified by means of the non-deterministic acceptor obtained by a composition of the transducer realising the reduction $J_1 \lhd J_2$, and the non-deterministic acceptor verifying that $J_2 \in NP$. The composition is performed in the same way as in the preceding proof. ●

Theorem 7.4.5. *Let* J_1 *and* J_2 *be languages and let* $J_1 \in P$. *If, moreover,* $\emptyset \neq J_2 \neq \{0, 1\}^*$, *then* $J_1 \lhd J_2$.

Proof. Choose words $w_+ \in J_2$ and $w_- \notin J_2$. We construct a reduction $J_1 \lhd J_2$ as follows: in the acceptor which proves that $J_1 \in P$ each instance of the instruction *ACCEPT* or *REJECT* is substituted by a jump to the sequence of introductions which will first write w_+ or w_-, respectively, onto the input tape, and then will halt the computation by the instruction *STOP*. ●

The main result of the present section is the following theorem due to S A Cook [75].

Theorem 7.4.6. *The problem of determination of satisfiability of logical formulae in conjunctive normal form is NP-complete.*

Proof. The fact that the above problem belongs to the class NP is easily proved by the following procedure: first assign non-deterministically the values true or false to the variables of the formula f under consideration and then determine the value of the formula f, which can be performed in time proportionate to the length of the formula. If we can obtain, in this way, the value $f = $ true, then the formula f is satisfiable.

The latter part of the proof of our theorem is essentially more difficult. Let $J \in NP$, and let M be a non-deterministic acceptor which accepts the language J in time bounded by the polynomial p. We will show how to construct for each word w a formula f_w in conjunctive normal form in such a way that f_w is satisfiable if, and only if, the acceptor M accepts w. If we prove that the construction can be performed in polynomial-bounded time, the theorem will be proved. We can assume that the polynomial p has the property mentioned in 2.2.3.

Denote by n the length of the word w, and put $m = p(n)$. By theorem 2.2.4 we can assume that the computation needs memory space at most m. Let q denote the number of instructions in the acceptor M. Each configuration of M obtained in the course of the first m steps of any admissible computation determined by the word w has the following properties.

1. The first n squares of the input tape are the only ones which can contain a number distinct from 0, and if the input head reads a square beyond the first n squares of the input tape, then its position does not influence the computation.

2. The first m memory registers are the only ones that can contain a number distinct from 0, and no memory register contains a member larger than m.

When trying to find out whether M accepts the word w, it is sufficient to know the first m steps of all admissible computations. In order to describe all those computations in that time segment completely, it is sufficient to know the values of the following system of logical variables:

a_i for $i = 1, ..., n$, where $a_i =$ true if the ith square of the input tape contains the number 1;

b_{it} for $i = 1, ..., n$ and $t = 0, ..., m$, where $b_{it} =$ true if the position of the input head is the ith square of the tape after the tth step of the computation;

c_{ijt} for $i = 0, ..., m$, $j = 0, \pm 1, ..., \pm m$ and $t = 0, ..., m$, where $c_{ijt} =$ true if the ith memory register contains the number j in the tth configuration of the computations; and

d_{jt} for $j = 1, ..., q$, and $t = 0, ..., m$, where $d_{jt} =$ true if the program counter contains the number j in the tth configuration of the computation.

The number of all the above variables is as follows:

$$n + n(m + 1) + (m + 1)^2 (2m + 1) + (m + 1) q = 0(p^3(n)),$$

and hence, their number is polynomial-bounded with respect to the number n.

To each admissible computation of the acceptor M we can assign values of the above variables. Conversely, if the above variables are to describe the performance of the first m steps of a computation of the acceptor M determined by the input word w and if the word is to be accepted within the time considered, then the following must hold:

1. the values of a_i correspond to the sequence of zeros and ones given by the word w;

2. for each t there exists at most one j such that $b_{it} =$ true (since the head cannot be at two distinct positions at one instant);

3. for each i and t there exists precisely one j such that $c_{ijt} = \text{true}$ (since each memory register contains precisely one number);

4. for each t there exists precisely one j such that $d_{jt} = \text{true}$;

5. the contents of the location counter in the mth step of computation which is determined by the values d_{jm} must yield the serial number of some of the instructions *ACCEPT*;

6. the correspondence of the values of the variables for given $t - 1$ and t must express a transition of the acceptor M from one configuration to another.

We now construct formulae $f_{w1}, ..., f_{w6}$ depending on the given variables in such a way that $f_{wi} = \text{true}$ holds if, and only if, the ith condition above is fulfilled.

1. The word w is a sequence of 0s and 1s. Choose as f_{w1} the formula

$$f_{w1} = A_1 \wedge A_2 \wedge ... \wedge A_n$$

where $A_i = a_i$ if the ith element of w is 1, else $A_i = \neg a_i$.

2. A formula $g(x_1, ..., x_n)$ which is satisfied if, and only if, at most one of the variables x_i takes the value true can be constructed as the conjunction of the expressions $\neg x_i \vee \neg x_j$, where i and j range over all pairs of numbers subject to $1 \leq i \leq j \leq r$. Given such a formula g, we can choose f_{w2} as follows:

$$f_{w2} = \bigwedge_{t=0}^{m} g(b_{1t}, b_{2t}, ..., b_{mt}).$$

3. Put

$$h(x_1, ..., x_r) = g(x_1, ..., x_r) \wedge (x_1 \vee x_2 \vee ... \vee x_r),$$

then $h = \text{true}$ if, and only if, $x_i = \text{true}$ holds for precisely one i. Using h, we can choose

$$f_{w3} = \bigwedge_{i=0}^{m} \bigwedge_{t=0}^{m} g(c_{i, -m, t}, ..., c_{i0t}, ..., c_{imt}).$$

4. Choose

$$f_{w4} = \bigwedge_{i=0}^{m} \bigwedge_{t=0}^{m} g(d_{-mt}, ..., d_{mt}).$$

5. f_{w5} is the disjunction of all those variables d_{jm} for which the jth instruction of the acceptor M is *ACCEPT*.

6. The construction of the last formula is the most difficult one because our task is, in fact, to describe the translation of the complete structure

of the acceptor M into a logic form, and moreover, each change of configuration is to be described individually. The formula f_{w6} will be the logical
product of formulae f_{w6r} for $r = 1, ..., q$, which express the role of the
individual instructions of the program. We are not going to describe all
the types of instructions, but only show one of the most difficult cases.

Assume that the rth instruction of the program is $ADD * 8$. Then we
require that the following hold for all $t = 1, 2, ..., m$:

If $d_{r,t-1} = $ true (i.e. if before performing the tth step of computation
the location counter contains the serial number of the instruction $ADD * 8$),
then for every x, y and z, if the variables $c_{0,x,t-1}, c_{1,y,t-1}$ and $c_{y+8,z,t-1}$ all
have the value true, that is, if in the $(t-1)$th configuration the accumulator
contains x, the index register contains y, and the memory register of the
address $y + 8$ contains z, then $c_{0,x+z,t}$ must also have value true, which
means that after performing the tth step of computation the contents of
the accumulator are increased by the contents of the memory register of the
address determined by the operand $*8$. Thus, the following formula must
be fulfilled:

$$\bigwedge_t \bigwedge_x \bigwedge_y \bigwedge_z (\neg d_{r,t-1} \vee \neg c_{0,x,t-1} \vee \neg c_{1,y,t-1} \vee \neg c_{y+8,z,t-1} \vee c_{0,x+z,t}).$$

Besides, the following formulae must be also fulfilled:

$$\bigwedge_{i=1}^{m} \bigwedge_{j=-m}^{m} \bigwedge_{t=1}^{m} k(c_{i,j,t-1}, c_{ijt}),$$

$$\bigwedge_{i=1}^{m} \bigwedge_{t=1}^{m} k(b_{i,t-1}, b_{it}),$$

$$\bigwedge_{j=1}^{q-1} \bigwedge_{t=1}^{m} k(d_{j,t-1}, d_{j+1,t}),$$

where $k(x, y) = (\neg x \vee y) \wedge (x \vee \neg y)$ is an auxiliary function satisfied if,
and only if, the values of x and y are equal. This serves to express the fact
that besides the change of the contents of the program counter and the
increase of the location counter by 1 no other change of configuration
takes place. The product of all the above formulae forms the formula f_{w6r}.

We suggest that the reader try to translate similarly at least one of the
instructions $LOAD$, $READ$, $ACCEPT$ and $CHOOSE$. For the last two,
the following implication: $d_{r,t-1} \Rightarrow (d_{L1,t} \vee d_{L2,t})$ must hold (besides
other conditions); in other words,

$$\neg d_{r,t-1} \vee d_{L1,t} \vee d_{L2,t}.$$

The resulting formula is in a conjunctive normal form, and it is not
difficult to verify that the whole construction can be performed in poly-

nomial-bounded time, although the degree of the polynomial is rather large. ●

The above theorem can be strengthened as follows.

Theorem 7.4.7. *The following problem is NP-complete.*
  Given a formula

$$f(x_1, ..., x_n) = (a_1 \lor b_1 \lor c_1) \land (a_2 \lor b_2 \lor c_2) \land ... \land (a_k \lor b_k \lor c_k),$$

where a_i, b_i and c_i are either variables or negations of variables $(i = 1, ..., k)$, decide whether f is satisfiable.

  Proof. The formula above is in the conjunctive normal form and hence the given problem is a special case of the problem of the preceding theorem. Consequently, the problem is a member of the class *NP*. Following theorems 7.4.3 and 7.4.6, it is thus sufficient to reduce the determination of satisfiability of general formulae in conjunctive normal form to the determination of satisfiability of the formulae of the special type under consideration. For this, it is sufficient to use the fact that the conjunction $u_1 \lor u_2 \lor ... \lor u_m$, $m > 3$, is satisfied if, and only if, for a suitable value of a newly adjoined variable v the following formula:

$$(u_1 \lor u_2 \lor ... \lor u_{m-2} \lor v) \land (u_{m-1} \lor u_m \lor \neg v)$$

is satisfied. In fact:
  if $u_1 \lor u_2 \lor ... \lor u_{m-2}$ is satisfied, it is sufficient to choose $v = $ false,
  if $u_{m-1} \lor u_{m-1}$ is satisfied, it is sufficient to choose $v = $ true,
and
  if $u_1 = ... = u_m = $ false, it is impossible to choose the value of v in such a way that the formula $(u_1 \lor u_2 \lor ... \lor u_{m-2} \lor v) \land (u_{m-1} \lor u_m \lor \neg v)$ is satisfied.

  By a successive application of the above procedure it is possible to transform an arbitrary formula in conjunctive normal form to the form in which no clause contains more than three literals. If necessary, the clauses with less than three summands can be completed by a repetition of the summands. ●

We will often use the following theorem.

Theorem 7.4.8. *The following assertions are equivalent.*
  1. **J** *is an NP-complete language.*
  2. **J** $\in NP$, *and there exists an NP-complete language* **K** *with* **K** $\lhd$ **J**.

Proof. If **J** is an NP-complete language, then $\mathbf{J} \in NP$, and in statement 2 we can choose **K** equal to **J**. Conversely, if statement 2 holds, then for each $\mathbf{L} \in NP$ we have, by definition 7.4.2, $\mathbf{L} \lhd \mathbf{K}$ and hence, by theorem 7.4.3, it follows that $\mathbf{L} \lhd \mathbf{J}$. ●

Theorem 7.4.8 is meaningful due to Cook's theorem above which ensures the existence of an NP-complete problem, and in fact describes one.

Applications to theorem 7.4.8 will much simplify proofs of NP-completeness, since it will only be necessary to reduce a unique problem to the given language, whereas definition 7.4.2 requires the reduction of all problems of the class NP. How are we going to proceed with such proofs? It is first necessary to prove that the problem under consideration belongs to the class NP. This is particularly easy when the problem has a formulation of the following type: 'Prove that there exists a set (or a permutation, a mapping, etc) which has a property V', provided that it is easily verifiable that V is satisfied. In that case, it is possible to construct the required non-deterministic algorithm accepting the given problem in such a way that the set (or other object) is first 'guessed', i.e. constructed in a non-deterministic manner, and then a standard deterministic procedure is applied to verify whether V is satisfied. For this reason, we shall often omit the proof that the problem under consideration belongs to the class NP.

In the second part of the proof of the NP-completeness it is necessary to choose a suitable NP-complete problem, and to reduce it to the problem under consideration. It is appropriate to choose the problem whose NP-completeness is the basis of our proof in such a way that it has features similar to those of the problem under consideration, and, primarily, to choose the simplest and most specific formulation because all the features of the chosen problem will have to be reproduced in the problem under study. Thus, for example, the problem of theorem 7.4.7 is more useful than the direct application of Cook's theorem because for the construction of reduction we can make use of the small number of summands in the individual factors.

It sometimes happens that we do not manage to prove that the problem **J** under consideration belongs to the class NP, although we are able to find the required reduction of some NP-complete problem to **J**. This may indicate that the problem **J** is in fact more difficult than any problem of the class NP. For those cases we introduce the following concept.

Definition 7.4.9. A language **J** is said to be *NP-hard* if for each $\mathbf{K} \in NP$ we have $\mathbf{K} \lhd \mathbf{J}$. ●

In order to prove that a problem **J** is *NP*-hard, it is sufficient to find a reduction **K** ◁ **J** where **K** is *NP*-complete or at least *NP*-hard.

The class of all *NP*-complete problems is sometimes denoted by *NPC*. We exhibit a simple property of this class.

Theorem 7.4.10. *If* $P \neq NP$, *then* $P \cap NPC = \emptyset$.

Proof. Suppose **J** $\in P \cap NPC$ and **K** $\in NP$. According to definition 7.4.2 we have **K** ◁ **J**, and by composing the algorithm which is used for this reduction with the deterministic algorithm ensuring that **J** $\in P$, we obtain an algorithm solving **K** deterministically in polynomial-bounded time. Thus, we see that $NP \subset P$ and it is sufficient to use theorem 7.3.6. ●

Since in this book we assume that most probably $P \neq NP$, · we have to admit that no *NP*-complete problem allows an algorithm working in polynomial-bounded time.

In the remaining section of the present chapter we are going to prove the *NP*-completeness of concrete problems from the theory of graphs, combinatorics, and operations research. It is important to realise that by proving the *NP*-completeness of a problem, we verify that it has a high complexity. Such a result is thus the more valuable, the simpler the problem seems to be at first sight. We will, consequently, investigate the possibility of weakening the given problem so as to preserve the *NP*-completeness.

7.5 Graph colouring

It is easy to determine whether a given graph can be coloured by k colours for $k = 1$ or 2. A graph can be coloured by one colour if, and only if, it has no edges, and by two colours if, and only if, it is bipartite. The latter can be verified in the case of connected graphs as follows: we colour an arbitrarily chosen vertex by colour number 1, all neighbours of that vertex by colour number 2, all the not yet coloured neighbours of those new vertices by colour 1, etc; such an alternating colouring is the only one possible, and if it does not yield the desired result, then the graph has a chromatic number at least 3. For unconnected graphs, this procedure is performed on each component individually. Consequently, using the methods of search given in Chapter 4, it is possible to determine the existence of a colouring by two colours in running time proportionate to the sum of the number of vertices and edges.

The boundary between solvability in polynomial-bounded time and *NP*-completeness is usually quite sharp. In the present problem, it turns out that already the question of existence of a colouring by three colours is NP-complete, and this is true even if we restrict ourselves to planar graphs with degrees of vertices smaller than or equal to 4. Those graphs can, however, always be coloured by four colours, which can be deduced in two distinct ways: the first one is by the planarity (see the successful solution of the problem of four colours mentioned earlier in Chapter 1), and the second is by the bound on the degrees. According to Brook's theorem, well known in graph theory, the only graph which has degrees of vertices smaller than or equal to 4 and which cannot be coloured by four colours is the complete graph on five vertices. The last graph is not planar, due to Kuratowski's theorem.

By restricting our attention to graphs with degrees smaller than or equal to 3, the colouring problem would become simple since the algorithm constructed according to the proof of Brook's theorem easily colours each such graph with three colours, except for the complete graph on four vertices.

The proof of the main result of the present section is divided into three theorems which stem from $[129]$.

Theorem 7.5.1. *The problem of determining whether a given graph can be coloured by three colours is NP-complete.*

Proof. It is easy to verify that the problem under consideration belongs to the class NP. For a reduction we choose the problem of theorem 7.4.7. We are going to colour graphs by three colours called 0, 1 and 2. By considering all possibilities we can verify that the auxiliary graph of figure

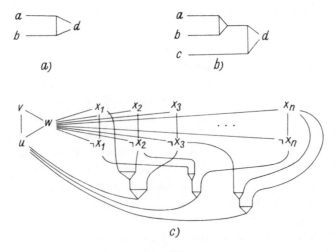

Figure 28.

28a has the following property with respect to colouring by three colours: if the vertices a and b have the same colour, then the vertex d must also be coloured by that colour, and if the colours of a and b are distinct, then d can be coloured arbitrarily. It follows that for the colouring of the graph of figure 28b by the colours 0, 1 and 2 the following holds: if the vertices a, b and c are all coloured by the colour 0, then the vertex d must also be coloured by 0, and if at least one of the vertices a, b and c is coloured by 1, then d can, or must, be coloured by 1.

Suppose we are given a formula

$$f(x_1, ..., x_n) = (a_1 \lor b_1 \lor c_1) \land ... \land (a_k \lor b_k \lor c_k)$$

where a_i, b_i and c_i are variables or negations of variables. The graph corresponding to that formula will be constructed in two parts as indicated by figure 28c. The basic skeleton is formed by the vertices u, v and w, which are pairwise connected, and by the vertices corresponding to variables and negations of variables. For each i, the vertices w, x_i and $\neg x_i$ are pairwise connected. For each clause of the formula f of the form $(a_i \lor b_i \lor c_i)$ we add a new copy of the graph in figure 28b connected by the vertices a, b and c to the vertices of the skeleton corresponding to a_i, b_i and c_i respectively. Moreover, the vertices d of those added graphs are connected with the vertex u. Figure 28c depicts the resulting graph of the following formula:

$$(x_1 \lor \neg x_2 \lor \neg x_3) \land (\neg x_1 \lor \neg x_2 \lor \neg x_n) \land ... \land (\neg x_3 \lor x_1 \lor x_n).$$

Without loss of generality, we can assume that the graph of figure 28c is coloured in such a way that the colours of the vertices u, v and w are 0, 1 and 2, respectively. Then for each i, there are two possibilities: the colours of x_i and $\neg x_i$ are either 0 and 1, respectively, or 1 and 0, respectively. This corresponds to the two possible values $x_i = $ false and $x_i = $ true. Moreover, the vertices d of the added graphs cannot have the colour 0 since they are connected with u. From the above we conclude that this can be fulfilled only if for each j at least one of the vertices a_j, b_j and c_j has the colour 1.

We have thus proved that the resulting graph can be coloured by three colours if, and only if, the formula f above is satisfiable. Observe that the constructed graphs can always be coloured by four colours. It is also evident that the construction can be performed in a polynomial-bounded time. ●

Theorem 7.5.2. *The problem of determining whether a given planar graph can be coloured by three colours is NP-complete.*

Proof. We will use the *NP*-completeness of the colouring problem described in the preceding theorem. We will colour again by the colours 0, 1 and 2. Consider the graph of figure 29a. We leave it to the reader to verify that the following table yields a colouring of that graph, and that all

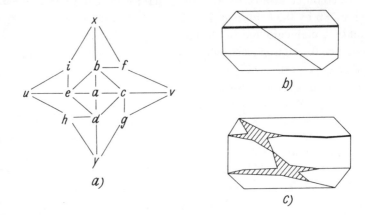

Figure 29.

colourings by the colours 0, 1 and 2 can be obtained from that colouring by a permutation of colours:

a	b	c	d	e	f	g	h	i	x	y	u	v
0	1	2	1	2	0	0	0	0	2	2	1	1
0	1	2	1	2	2	1	2	1	0	0	0	0

We conclude that the vertices x and y must have the same colour, and the same is true for the vertices u and v, but the colours of x and u are independent.

Let a graph G be given. We choose a drawing of the graph in the plane; if the graph is not planar, there necessarily arise crossings of the edges, see figure 29b. If we substitute those crossings by gluing onto them the four-corner graph of figure 29a in the manner indicated by figure 29c, we obtain a graph which can be coloured by three colours if, and only if, the same is true about the original graph G. To verify this, let us follow the bold edge of figure 29b. The corresponding part of the graph of figure 29c is the four-corner crossing graph transporting the colour of the left end of the line to its right corner which is also the left end-vertex of the shorter bold edge of figure 29c substituting the edge of figure 29b under consideration.

It is evident that the graph constructed in this way is planar, and that the above construction can be performed in polynomial-bounded time. ●

Theorem 7.5.3. *The problem of determining whether a given planar graph with degrees of vertices smaller or equal to 4 can be coloured by three colours is NP-complete.*

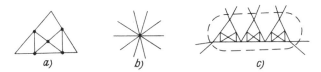

Figure 30.

Proof. We leave it to the reader to verify that each colouring of the triangle graph of figure 30a by three colours gives the same colour to all the three vertices determining the triangle. Given a planar graph, we will substitute each vertex of degree larger than 4 by a formation composed of copies of the graph of figure 30a in the manner indicated by figures 30b and c. We will obtain a planar graph which can be coloured by the three colours if, and only if, the same is true for the original graph. The degrees in the new graph do not exceed 4. It is clear that the construction can be performed in polynomial-bounded time.

7.6 Searching an independent set

The following assertion follows immediately from the construction presented in example 7.1.2 and from Cook's theorem.

Theorem 7.6.1. *The problem of determining whether a given graph contains an independent set of a prescribed size is NP-complete.* ●

We will try to simplify the problem as much as possible while preserving the *NP*-completeness: we will show that it is possible to restrict our attention to planar graphs with vertices of degree at most 3. The results of the present section originate from [129]. Observe that by restricting

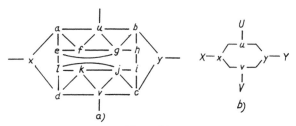

Figure 31.

the degrees of vertices to at most 2, we would get a simply solvable problem since in such a graph, the components are formed by paths and circuits.

We are going to make use of the properties of the auxiliary graph of figure 31a.

Lemma 7.6.2. *Let Z be an independent set of the graph of figure 31a. By the type of the set Z we will understand the following triple (a_1, a_2, a_3) of numbers:*

a_1 *is 1 or 0 according to whether the vertex x does or does not lie in Z;*

a_2 *is 1 or 0 according to whether the vertex y does or does not lie in Z; and*

$a_3 = 0, 1$ *or 2 is the number of elements of the set $\{u, v\}$ lying in Z.*

Then the following hold:

1. *every independent set Z has at most six elements,*

2. *there exist independent six-element sets of both types $(1, 0, 1)$ and $(0, 1, 1)$,*

3. *all sets of the types $(0, 0, 1), (0, 1, 0), (1, 0, 0), (0, 0, 2)$ and $(1, 1, 0)$ have at most five elements,*

4. *every set of type $(0, 0, 0)$ has at most four elements.*

Proof. The basic observation is that an odd circuit having $2k + 1$ elements can meet the set Z in at most k elements.

1. The circuits $x, a, e, l, d,$ and y, b, h, i, c both contain at most two elements of the set Z, and the circuits $u, f, g,$ and v, j, k both contain at most one element of Z.

2. Consider $\{x, u, c, e, h, k\}$ and $\{y, u, d, e, h, j\}$.

3. and 4. Each of the circuits $a, e, f; b, h, g; c, i, j; d, l, k$ contains at most one element of the set Z, and therefore, for the types $(0, 0, 1), (0, 1, 0),$ and $(1, 0, 0)$, the number of elements cannot exceed five and for the type $(0, 0, 0)$, the number is at most four. For the type $(0, 0, 2)$ we know that u and v lie in Z and hence further elements of Z can only be as follows: one of the vertices e and l, and one of the vertices h and i — thus, we have at most four elements. Finally, for the type $(1, 1, 0)$ we have $x, y \in Z$ and $u, v \in Z$ and hence further elements of Z must lie in the interior of the graph of figure 31a, i.e. at most one of the vertices e, f and g, at most one of j, k and l, and at most one of h and i. ●

Theorem 7.6.3. *The problem of determining whether a given planar graph has an independent set of a prescribed size is NP-complete.*

Proof. We proceed as in the proof of theorem 7.5.2. We draw the given graph in the plane, and we substitute each crossing of edges by the graph of figure 31a in the manner indicated by figure 31b, where the original crossing edges were X, Y and U, V. In contrast to 7.5.2, the graph we are gluing does not meet the original vertices, nor do those graphs meet one another.

If m crossings are substituted in the above way, then the original graph G has an independent set of k elements if, and only if, the new graph has an independent set of $k + 6m$ elements. This is seen from the following facts.

If the original graph has an independent set Z of k elements, then for each crossing as in figure 31b, the set Z can be expanded by six vertices of the new graph, forming an independent set of type either $(1, 0, 1)$ or $(0, 1, 1)$, because we cannot have both $X \in Z$ and $Y \in Z$, and we cannot have both $U \in Z$ and $V \in Z$.

Conversely, if Z_1 is an independent set of the new graph, then in the situation indicated by figure 31b we have the following possibilities: either $X, Y \in Z_1$ or $U, V \in Z_1$ or even $X, Y, U, V \in Z_1$, and then Z_1 determines in the auxiliary graph an independent set of the type $(0, 0, ?)$, $(?, ?, 0)$ or $(0, 0, 0)$, respectively. The latter set has, according to the preceding lemma, at most five elements, and in the last case at most four elements. We can thus perform the following operation: if $X, Y \in Z_1$, we delete the vertex Y from Z_1, and if $U, V \in Z_1$, we delete the vertex U from Z_1. Then we change the form of Z_1 in the auxiliary crossing graph to a six-element set of type $(0, 1, 1)$ (which can be performed preserving the independence by assertion 2 of lemma 7.6.2). We thus add to Z_1 at least the same number of elements we have deleted from it. This procedure will change the set Z_1 in such a way that its size will not be decreased and that by deleting at most $6m$ elements lying in the auxiliary graphs we have glued into the crossing we will obtain an independent set of the original graph. ●

Theorem 7.6.4. *The following problem is NP-complete: determine whether a given planar graph whose vertices have degrees smaller than or equal to three has an independent set of a prescribed size.*

Proof. If we substitute, in a given planar graph, vertices of degrees $d > 3$ by an auxiliary graph connected to the neighbours of the original

Figure 32.

vertex in the manner indicated by figures 32a and b, we obtain a planar graph which has no vertices of degree larger than 3. We will prove that one such substitution will increase the size of the largest independent set precisely by $d - 1$.

In the circuit of $2d$ vertices which forms the basis of the auxiliary graph we can choose the largest independent set of d elements in two ways: we choose either the vertices connected to the neighbouring vertices, or the remaining ones, indicated in the figure by black dots. In the latter case, the additional edge will force the deletion of one of its end-vertices, provided that the mentioned set is to remain independent in the whole auxiliary graph. We now have two possibilities:

either the vertex under substitution lies in the independent set of the original graph, in which case none of its neighbours lies in the set, and then instead of the original vertex we can add d vertices of the auxiliary graph to the independent set,

or the vertex under substitution does not lie in the independent set and then, assuming that the set was of maximum size, the set must contain one of the neighbours of the vertex, in which case we can use $d - 1$ vertices of the auxiliary graph, but not d vertices.

By successive substitution of the vertices with large degrees we will eventually obtain, in polynomial-bounded time, a graph with the required properties, and we will know precisely the correspondence between the size of independent sets in the original graph and in the new one. ●

7.7 Coverings of sets

By a *covering* of a set X is understood a collection of subsets of X, the union of which is equal to X. Covering can be understood as a generalisation of the concept of a graph without isolated vertices, i.e. without vertices which are not end-vertices of any edge. (For example, each connected graph of at least two vertices has that property.) In fact, according to our definition, edges of a graph are two-point subsets of the set of vertices of the graph. We will show that the problem of finding a maximum matching, which we studied in §6.4, becomes, when generalised to a system of three-element subsets, NP-complete. Moreover, this complexity remains even if we restrict ourselves to a certain generalisation of the problem of a perfect matching (i.e. a matching without free vertices) in a bipartite graph with bounded degrees of vertices, which we describe below; see [168].

Theorem 7.7.1. *The following problem is NP-complete:*
There are given three disjoint sets X_0, X_1 and X_2 of the same size and a covering $\mathscr{S}$ of the set $X = X_0 \cup X_1 \cup X_2$ with the following properties:
(a) each set of the covering $\mathscr{S}$ meets each of the sets X_0, X_1 and X_2 in precisely one element (and has, therefore, three elements),

(b) *for each element x of the set X there exist precisely three subsets lying in the covering $\mathscr{S}$ which contain x.*

Determine whether $\mathscr{S}$ contains a disjoint decomposition of the set X, i.e. whether there exists $\mathscr{S}_0 \subset \mathscr{S}$ such that for each $x \in X$ there exists precisely one $s \in \mathscr{S}_0$ with $x \in s$.

Proof. We apologise to the reader for having chosen a less graphic construction in order to shorten the proof. We will use the fact that the problem of colouring of a graph by three colours is NP-complete, as shown in 7.5.1.

Let $G = (V, H)$ be a graph. Put

$$Z = \{(v, h) \mid v \in V,\ h \in H \text{ and } v \text{ is an end-vertex of } h\}.$$

Without loss of generality we can assume that the sets V, H, and Z are pairwise disjoint. Put

$$W = V \cup H \cup Z,$$
$$X = W \times \{0, 1, 2\} \times \{0, 1, 2\},$$
$$X_i = W \times \{0, 1, 2\} \times \{i\}, \qquad i = 0, 1, 2.$$

We now define $\mathscr{S}$ as a collection of triples. We make use of a mapping p constructed in the following way: for each vertex v of the graph G we order the edges incident on v in an arbitrary way into a sequence $h_1, \ldots, h_j$, and we put

$$p(v) = (v, h_1),\ p((v, h_1)) = (v, h_2),\ \ldots,\ p((v, h_{j-1}))$$
$$= (v, h_j),\ p((v, h_j)) = v.$$

The sets of the collection $\mathscr{S}$ are the following ones:

$$\{(v, 0, 0), (v, i, 1), (v, i, 2)\}\ v \in V, \qquad i = 0, 1, 2,$$
$$\{(h, i, 0), ((v, h), i, 1), ((v, h), i, 2)\}, (v, h) \in Z, \qquad i = 0, 1, 2,$$
$$\{(h, i, 0), (h, i, 1), (h, i, 2)\}, h \in H, \qquad i = 0, 1, 2,$$
$$\{(v, j, 0), (v, i, 1), (p(v), i, 2)\}, v \in V, \qquad i = 0, 1, 2,\ j = 1, 2,$$
$$\{((v, h), j, 0), ((v, h), i, 1), (p((v, h)), i, 2)\}, (v, h) \in Z, \qquad i = 0, 1, 2,\ j = 1, 2,$$
$$\{((v, h), 0, 0),\ (h, i, 1), (h, i, 2)\}, (v, h) \in Z, \qquad i = 0, 1, 2.$$

The property (a) above is evidently fulfilled. We leave it to the reader to check property (b) and the fact that given a colouring f of the vertices of the graph G by the colours $0, 1$ and 2, then the following system is a disjoint decomposition contained in $\mathscr{S}$:

$$\{(v, 0, 0), (v, f(v), 1), (v, f(v), 2)\}, v \in V,$$
$$\{(h, f(v), 0), ((v, h), f(v), 1), ((v, h), f(v), 2)\}, (v, h) \in Z,$$
$$\{(h, i, 0), (h, i, 1), (h, i, 2)\}, h \in H,$$

where i is the (unique) colour which has not been used to colour any of the end-vertices of the edge h,

$$\{(v, j, 0), (v, f(v) + j \bmod 3, 1), (p(v), f(v) + j \bmod 3, 2)\}, v \in V, \qquad j = 1, 2,$$
$$\{((v, h), j, 0), ((v, h), f(v) + j \bmod 3, 1), (p(v, h), f(v) + j \bmod 3, 2)\},$$
$$(v, h) \in Z, \qquad j = 1, 2,$$
$$\{((v, h), 0, 0), (h, f(v), 2)\}, (v, h) \in Z.$$

Here $i + j \bmod 3$ denotes whichever of the numbers $i + j$ or $i + j - 3$ lies in the set $\{0, 1, 2\}$.

The verification of the required properties is a lengthy but evidently mechanical task.

Assume, conversely, that a disjoint decomposition $\mathscr{S}_0$ has been chosen from $\mathscr{S}$. Then for each $v \in V$ there must exist exactly one number of the set $\{0, 1, 2\}$, to be denoted by $f(v)$, such that

(c) $$\{(v, 0, 0), (v, f(v), 1), (v, f(v), 2)\} \in \mathscr{S}_0 .$$

It is our task to show that the mapping f defined in this way is a colouring of the graph G by the colours 0, 1 and 2. To this end we will show that each $(v, h) \in Z$ satisfies the following:

(d) $$\{(h, f(v), 0), ((v, h), f(v), 1), ((v, h), f(v)' 2)\} \in \mathscr{S}_0 .$$

Suppose that, on the contrary, (v, h) is an element of Z which does not satisfy (d). Assume that the edges incident on v are ordered as $h_1, ..., h_j$ with $h = h_k$ for a suitable k, and that

$$p(v) = (v, h_1), p((v, h_i)) = (v, h_{i+1}) \qquad \text{for } i = 1, ..., j - 1.$$

Without loss of generality we can assume that (d) holds for all pairs (v, h_i) with $i < k$, since we would else choose as (v, h) a pair (v, h_i) with a smaller value of the index i.

The only three elements of the covering $\mathscr{S}$ containing $((v, h), f(v), 2)$ are the following:

$$\{(h, f(v), 0), ((v, h), f(v), 1), ((v, h), f(v), 2)\} \quad \text{and}$$
$$\{(a, j, 0), (a, f(v), 1), ((v, h), f(v), 2)\}, \quad j = 1, 2,$$

where $a = v$ if $k = 1$, or $a = (v, h_{k-1})$ if $k > 1$.

Since (d) does not hold, at least one of the last two of the above sets must lie in $\mathscr{S}_0$. This is a contradiction because either $a = v$, and then we conclude from (c) that

$$\{(v, 0, 0), (v, f(v), 1), (v, f(v), 2)\} \in \mathscr{S}_0 ,$$

or $a = (v, h_{k-1})$ and since (d) holds for $((v, h_{k-1}), f(v), 2)$, we conclude

$$\{(h, f(v), 0), ((v, h_{k-1}), f(v), 1), ((v, h_{k-1}), f(v), 2)\} \in \mathcal{S}_0 .$$

Suppose now that an edge h of the graph G is given with end-vertices v_1 and v_2. Then

$$\{(h, f(v_i), 0), ((v_i, h), f(v_i), 1), ((v_i, h), f(v_i), 2)\} \in \mathcal{S}_0 \quad \text{for } i = 1, 2,$$

and hence, $f(v_1) \neq f(v_2)$ else the element

$$(h, f(v_1), 0) = (h, f(v_2), 0)$$

would be covered by two distinct sets.

We have thus proved that the graph G can be coloured by three colours if, and only if, a disjoint decomposition can be chosen from $\mathcal{S}$. It is not difficult to check that the construction of X and $\mathcal{S}$ above can be performed in polynomial-bounded time. ●

We have thus shown that the choice of a disjoint decomposition from a covering formed by three-element sets is a difficult problem. The special form of the problem of covering makes it possible to conclude immediately the NP-completeness of another covering problem.

Theorem 7.7.2. *The problem of determining whether for a given natural number k and a given covering $\mathcal{S}$ of a set X there exists a covering $\mathcal{S}_0$ of the set X containing k subsets of X and such that $\mathcal{S}_0 \subset \mathcal{S}$ is NP-complete.*

Proof. This is an immediate consequence of theorem 7.7.1, since a covering $\mathcal{S}_0$ chosen from a covering $\mathcal{S}$ of a set X by three-element subsets is a disjoint decomposition if, and only if, it has $N/3$ elements, where N is the number of elements of the set X. ●

7.8 Integer problems

In this section we shall prove that the problem of integer linear programming is NP-complete, and the same is true of very special cases of the problem. By comparing this result with the existence of an algorithm for the problem of linear programming with real values of variables for which an algorithm working in polynomial-bounded time has, in fact, been found (see [69]), we conclude that the requirement of integers in admissible solutions makes the complexity of the problem substantially higher. On the other hand, it is necessary to remark that there exist fast approximate algorithms which solve many of the problems of the present section, for example the knapsack problem, within an arbitrarily prescribed precision, see [168].

In order to avoid misunderstandings which are often encountered in interpretations of the actual meaning of the NP-completeness of integer problems, we have to be very careful. The question of the existence of an algorithm solving a given problem in polynomial-bounded time depends on the way we determine the size of the input data. When formally describing our problem as a language, every set of input data is understood as a sequence of zeros and ones, and there are no difficulties concerning the interpretation of size. On the other hand, for problems presented verbally one often uses a number of different ways of evaluating the size. For example, the size of a graph is usually understood as the number of its vertices, but if we describe the graph as a sequence of zeros and ones, for example using the incidence matrix, we need n^2 entries. If an algorithm processing the graph works in running time bounded by a polynomial $p = p(N)$ depending on the size of the problem determined by the number, $N = n^2$, of bits of the incidence matrix, then it also works in running time bounded by the polynomial $q(n) = p(n^2)$ of double the degree depending on the number n of vertices of the graph. If we are only interested in the question of existence of an algorithm working in polynomial-bounded time, then it is irrelevant whether the size of a graph is interpreted as the number of vertices, or the sum of the numbers of vertices and edges, or the number of coefficients of the incidence matrix.

In the case of integer problems the question of determination of the size of input data is very subtle. Let $a_1, ..., a_n$ be a sequence of integers. We often interpret the size of the sequence as the maximum of the number n and the absolute values of the numbers $a_1, ..., a_n$. We will call this interpretation 'practical', and we will also use a different interpretation: the size of the sequence is the sum of the bits necessary for a presentation of the numbers a_i, which is approximately $\sum \log |a_i|$. Both ways are related in the common situations where each of the numbers a_i requires approximately $\log n$ bits.

A sequence of size N in the practical interpretation cannot contain a number larger than N, while in the logarithmic interpretation it can even contain numbers close to 2^N. By methods of dynamic programming it is possible, for the problems investigated in the present section, to construct algorithms working in running time bounded by a polynomial which depends on the size of the input data determined in the practical way. On the other hand, we are going to present results on NP-completeness which imply that it is probably impossible to construct algorithms for finding the optimum solution in running time bounded by a polynomial dependent on the size of the input data in the logarithmic interpretation.

In other words, we are able to process quite quickly input data containing a large number of integers of orders of hundreds and thousands, since this is roughly the usual number of iterations of the algorithms of dynamic programming. But we are totally incapable of searching the exact solution whenever we are to process numbers with tens and hundreds of binary or decadic digits. However, such numbers are encountered only rarely, which often leads to a misinterpretation of the results presented below.

We first present an example illustrating the different relationships of the real and integer variant of the problem of linear programming to *NP*-completeness.

Example 7.8.1. Let the following formula of the type considered in theorem 7.4.7 be given:

$$(a_{11} \lor a_{12} \lor a_{13}) \land \ldots \land (a_{m1} \lor a_{m2} \lor a_{m3})$$

with variables $x_1, \ldots, x_n$. Let $X_1, \ldots, X_n$, $S_1, \ldots, S_m$ be non-negative real variables satisfying the following inequalities

$$X_k \leq 1, \ S_i \leq 1 \quad \text{for} \quad k = 1, \ldots, n, \quad i = 1, \ldots, m,$$

$$S_i - (A_{i1} + A_{i2} + A_{i3}) \leq 0, \quad \text{for} \quad i = 1, \ldots, m,$$

where $A_{ij} = X_k$ if $a_{ij} = x_k$, and $A_{ij} = 1 - X_k$ if $a_{ij} = \neg x_k$.

It is our task to decide whether the maximum value of the sum $S_1 + \ldots + S_m$ is m (it cannot be larger).

If the variables above are to have integer values, then they are equal to 0 or 1, and S_i can be equal to 1 only if $A_{ij} = 1$ for at least one j. The question whether the above maximum sum is equal to m is thus equivalent to the question of satisfiability of the original formula. If, on the other hand, we admit real values of the variables, we can put $X_1 = \ldots = X_k = 0.5$, which makes it possible that each S_i be equal to 1, but which has no interpretation in classical logic. ●

Example 7.8.1 is actually a proof of the *NP*-completeness of the decision variant of the problem of integer programming. This assertion can be substantially strengthened.

Theorem 7.8.2. [139]. *The problem of determining whether for given natural numbers $n_0, \ldots, n_k$ and N there exists a set $A \subset \{1, \ldots, k\}$ with*

$$\sum_{i \in A} n_i = N$$

is NP-complete.

Proof. We are going to perform a reduction of the problem described in theorem 7.7.1. Let $\mathscr{S} = \{s_1, \ldots, s_k\}$ be a covering of the set $X = \{x_1, \ldots, x_N\}$. Put $d = n + 1$. Given non-negative integers a_i and b_i for $i = 0, \ldots, k$, which are smaller than d, then the following equality

$$a_k d^k + \ldots + a_0 d^0 = b_k d^k + \ldots + b_0 d^0$$

implies $a_i = b_i$ for $i = 0, \ldots, k$, since a_i are actually the digits of the d-ary expansion of the number $a_k d^k + \ldots + a_0 d^0$, and analogously with b_i.

We now put $F(x_i) = d^{i-1}$ for $i = 1, \ldots, n$, and we evaluate each subset Y of the set X by the following number:

$$F(Y) = \sum_{x \in Y} F(x).$$

If a disjoint decomposition of the set X can be chosen from $\mathscr{S}$, then it is possible to choose a subset of the set of numbers $F(s_1), \ldots, F(s_k)$ the sum of which is $F(X)$. In fact, it is sufficient to choose the numbers $F(s_i)$ corresponding to the sets s_i of the disjoint decomposition. We prove the converse: suppose $\mathscr{S}_0 \subset \mathscr{S}$. Then we have

$$\sum_{s \in \mathscr{S}_0} F(s) = \sum_{s \in \mathscr{S}_0} \sum_{x \in s} F(x) = \sum_{x \in X} \sum_{x \in s \in \mathscr{S}_0} F(x)$$

$$= \sum_{x \in s} u_x F(x) = \sum_{i=1}^{n} u_{x_i} d^{i-1}, \qquad F(X) = \sum_{i=1}^{n} 1 \cdot d^{i-1},$$

where u_x denotes the number of elements of the collection $\mathscr{S}_0$ containing x. It follows from the equality of the above expressions and from the inequality $u_x \leq n < d$ that the above statement implies $u_x = 1$ for each $x \in X$, which means that $\mathscr{S}_0$ is a disjoint decomposition of X.

All the numbers $F(s_i)$ and $F(X)$ can be expressed in binary system by means of at most $(n + 1) \log (d + 1)$ digits 0 and 1, since they do not exceed the number $1 + \ldots + d^{n-1} = d^n - 1$, and hence, a description of the sequence of numbers constructed above requires only $O(nm \log n)$ bits. This means that the whole construction can be performed in polynomial-bounded time. ●

7.9 The travelling salesman problem

The well-known travelling salesman problem is formulated as follows: a set M is given and for each pair x, y of members of X a number $d(x, y)$ is specified which we will call the distance of x and y. For simplicity, we will assume that d is a metric, i.e. the following hold:

$d(x, y) \geqq 0$ for arbitrary x and y with equality if, and only if, $x = y$.

$d(x, y) = d(y, x)$ for arbitrary x and y,

$d(x, y) + d(y, z) \geqq d(x, z)$ for arbitrary x, y and z (the triangle inequality).

It is our task to determine in which ordering the travelling salesman should traverse the elements of the set M (called towns) if he wants to traverse each of them precisely once before returning to the town from which his journey was started, and if the distance he covers is to be the minimum possible one.

In other words, we are looking for an ordering of the elements of the set M into a sequence $x_1, ..., x_n$ in which each element of M is contained precisely once and such that the following sum

$$d(x_1, x_2) + d(x_2, x_3) + ... + d(x_{n-1}, x_n) + d(x_n, x_1)$$

has the minimum possible value.

A special case of the above problem is the search for a hamiltonian circuit in a graph: it is sufficient to put $M = V(G)$ and

$$d(x, y) = \begin{cases} 1 & \text{if } \{x, y\} \in E(G) \\ 2 & \text{if } \{x, y\} \notin E(G) \end{cases}$$

and then to ask whether there exists a path of the travelling salesman of length equal to the number of vertices of the graph G.

The question of the existence of a hamiltonian cycle of a directed graph is a quite similar situation. The NP-completeness of all three of these problems has been shown by R M Karp in his paper [168].

Theorem 7.9.1. *The following three problems are NP-complete:*
  (a) *determine whether a given directed graph has a hamiltonian cycle;*
  (b) *determine whether a given graph has a hamiltonian circuit;*
  (c) *given a set M of n elements, a metric d on the set M and a number K, determine whether there exists a sequence $x_1, ..., x_n$ containing each element of M exactly once and such that*

$$d(x_1, x_2) + ... + d(x_{n-1}, x_n) + d(x_n, x_1) \leqq K.$$

Proof. The *NP*-completeness of problem (a) will be proved by means of theorem 7.6.1. Let G be a graph of n vertices. We are interested to know whether there exists an independent set of vertices of G having m elements.

For each vertex v of G we order the edges incident with v arbitrarily into a sequence $E_{v1}, ..., E_{vd}$ (where $d = \deg_G(v)$). Each edge $\{v, w\}$ thus appears twice: once as E_{vi} and once as E_{wj} for suitable i and j.

We now construct a directed graph H as follows:

the vertices of the graph H are triples (v, e, i) where v is a vertex of the graph G, e is an edge of G which is incident with v, and i is either 0 or 1; we furthermore add $n - m$ distinct vertices denoted by $a_1, ..., a_{n-m}$;

the edges of H are the following pairs:

$((v, e, 0), (v, e, 1)), v \in e \in E(G)$;

$((v, e, i), (w, e, i))$,   where $e = \{v, w\} \in E(G)$, $i = 0, 1$,

$((v, E_{vj}, 1), (v, E_{v,j+1}, 0)), v \in V(G), 1 \leq j < \deg(v)$,

$(a_p, (v, E_{vi}, 0)), 1 \leq p \leq n - m, v \in V(G)$,

$((v, E_{vd}, 1), a_p), 1 \leq p \leq n - m, v \in V(G), d = \deg(v)$.

Assume first that X is an independent set of the graph G of m vertices. We will show how to construct, under this assumption, a hamiltonian cycle in H.

Denote by Y the set of all vertices of G which do not lie in X. The number of elements of Y is $n - m$, and each vertex of the graph G has one end-vertex lying in Y. Denote by $y_1, ..., y_{n-m}$ the elements of the set Y and consider the following cycle of the graph H:

$$a_1, \quad (y_1, E_{y_1,1}, 0), (y_1, E_{y_1}, 1), ..., \quad (y_1, E_{y_1,\deg(y_1)}, 0), (y_1, E_{y_1,\deg(y_1)}, 1),$$
$$a_2, \quad (y_2, E_{y_2,1}, 0), \qquad\qquad ..., \quad (y_2, E_{y_2,\deg(y_2)}, 1),$$
$$\vdots$$
$$a_{n-m}, (y_{n-m}, E_{y_{n-m},1}, 0), \qquad ..., \quad (y_{n-m}, E_{y_{n-m},\deg(n-m)}, 1).$$

This cycle contains all vertices of the graph H except those of the form (v, e, i) for $v \in X$. Suppose $e = \{v, w\}$ is an edge of the graph G such that $v \in X$. Then, necessarily, we have $w \in Y$ and we can add to the above cycle the vertices $(v, e, 0)$ and $(v, e, 1)$ substituting at the same time the edge $((w, e, 0) (w, e, 1))$ by the following edges:

$$((w, e, 0), (v, e, 0)), ((v, e, 0), (v, e, 1)), ((v, e, 1), (w, e, 1)),$$

thus creating, eventually, a hamiltonian cycle of the graph H.

Conversely, suppose that a hamiltonian cycle of the graph H is given. If we delete the vertices $a_1, ..., a_{n-m}$ from it, we obtain $n - m$ directed paths. Each of the paths starts in a vertex of the form $(v, E_{v1}, 0)$, and we then call v the leading vertex of the corresponding path. The cycle then continues, for each $i = 1, 2, ..., \deg(v) - 1$, in one of the following ways:

either directly to $(v, E_{vi}, 1)$,

or to the vertex $(w, E_{vi}, 0)$ where w is the other end-vertex of the edge E_{vi}, and then it necessarily continues to $(w, E_{vi}, 1)$ and further necessarily to $(v, E_{vi}, 1)$, since to the last vertex there lead directed edges solely from the vertices $(v, E_{vi}, 0)$ and $(w, E_{vi}, 1)$.

The part of the hamiltonian cycle determined by the vertex v thus traverses successively the vertices (v, E_{vi}, j) for $i = 1, ..., \deg(v)$ and $j = 0, 1$ with eventual side-steps through the pairs $(w, E_{vi}, 0), (w, E_{vi}, 1)$. The number of leading vertices is $n - m$, and since the cycle under consideration is hamiltonian, each edge of the original graph G has a leading end-vertex. Consequently, the set of all vertices of G which are not leading ones forms an independent set of the graph G of m vertices. This proves the NP-completeness of problem (a).

The NP-completeness of problem (b) is seen by a simple reduction of problem (a): given a directed graph H, we construct a graph P whose vertices are the pairs (v, i) where $v \in V(H)$ and $i = 0, 1, 2$ and whose edges are the following pairs:

$$\{(v, 0), (v, 1)\}, \quad v \in V(H),$$
$$\{(v, 1), (v, 2)\}, \quad v \in V(H),$$
$$\{(v, 2), (w, 0)\}, \quad (v, w) \in E(H).$$

It is not difficult to verify that each circuit of the graph P has the following form:

$$(v_1, 0), (v_1, 1), (v_1, 2), (v_2, 0), (v_2, 1), (v_2, 2), ..., (v_k, 0), (v_k, 1), (v_k, 2),$$

where $v_1, v_2, ..., v_k$ is a cycle of H, and such a circuit is hamiltonian if, and only if, $v_1, ..., v_k$ is a hamiltonian cycle.

The NP-completeness of problem (c) follows immediately from the NP-completeness of problem (b), which is a special case of (c). ●

The problems considered in theorem 7.9.1 remain NP-complete even if we restrict ourselves to very special metrics or graphs. In order to keep our book concise, we present the next results without proof.

In the first place, it is possible to prove that the problem of determining whether a given planar 3-connected cubic graph contains a hamiltonian circuit is NP-complete, see [130]. (A graph is said to be *cubic* if each of its vertices has degree 3.) The history of the investigation of those graphs is very interesting: in 1880, P G Tait formulated a hypothesis that each planar 3-connected cubic graph contains a hamiltonian circuit. Although it has been known that the affirmative solution of Tait's hypothesis would solve the four-colour problem affirmatively, the question remained open until 1946 when it was answered in the negative by W T Tutte. And today it turns out that the determination of the existence of hamiltonian circuits in the above graphs is very difficult indeed.

The travelling salesman problem is very often investigated under the following restriction: the elements of M are points in the plane, and the distance of the points u and v of coordinates u_1, u_2 and v_1, v_2, respectively, is determined in one of the following ways:

$$d(u, v) = \max \left(|u_1 - v_1|, |u_2 - v_2| \right) \text{ (maximum metric)},$$
$$d(u, v) = |u_1 - v_1| + |u_2 - v_2| \text{ (sum metric)},$$
$$d(u, v) = \sqrt{\left[(u_1 - v_1)^2 + (u_2 - v_2)^2 \right]} \text{ (euclidean metric)}.$$

In [128 and 215] it is proved that if the elements of the set M are points in the plane with integer coordinates, and if we ask whether there exists a path of the travelling salesman which is not longer than a given integer K, then when d is the maximum metric or sum metric, the problem is NP-complete.

For the euclidean metric it has only been proved that the problem is NP-hard. It is, in fact, not clear whether the euclidean variant belongs to the class NP because it is not known whether it is possible to decide in polynomial-bounded time whether the following inequality:

$$\sqrt{m_1} + \sqrt{m_2} + \ldots + \sqrt{m_n} \leqq K$$

is satisfied for given natural numbers $m_1, \ldots, m_n$ and K (in the case when the time bound is related to the number of digits needed to write down the data). Direct methods of solution of the above inequality (by squaring, etc) are usually impractical, and a numerical computation makes it possible in a majority of cases to determine easily the validity or non-validity of the inequality, but since the difference between the left-hand side and the right-hand side can be arbitrarily small, it is impossible to choose in advance such a bound on the numerical precision of the computation of square roots which would ensure that the computation will be fast in all cases.

If, however, we change the definition of the metric d by choosing a number $\varepsilon > 0$ and defining $d(u, v)$ as the euclidean metric rounded off upwards to the nearest integer multiple of ε, then the difficulties with square roots are overcome, and this rounded-off variant is also known to be NP-complete.

7.10 Further *NP*-complete problems

Karp's paper [168] introducing the concept of *NP*-completeness and presenting a list of the most basic problems of that type has initiated an investigation of this field of problems to such an extent that in the 1970s, *NP*-completeness became one of the most fundamental directions of the theory of computational complexity.

The monograph [27], devoted to that field of problems, presented more than 300 *NP*-complete problems (not counting all their different modifications and special cases), and this number has been increasing ever since. An excellent survey to the most recent results in *NP*-completeness and related fields is a column of D S Johnson [162], which is published regularly in *Journal of Algorithms*.

In the present section, we exhibit some of the more interesting *NP*-complete problems, and in the exercises we present some simple problems, the proof of the *NP*-completeness of which can be left to the reader.

The following problems have been proved to be *NP*-complete.

1. Width of a graph [214]

Input data: a graph $G = (V, E)$ and natural number K.
Question: does there exist a labelling F of the vertices of G by the numbers $1, ..., n$, where n is the number of vertices of the graph G, such that for each edge $\{u, v\} \in E$ we have

$$|F(u) - F(v)| \leq K?$$

2. Optimum linearisation of a graph [129]

Input data: a graph $G = (V, E)$ and a natural number K.
Question: does there exist a labelling F of the vertices of G by the numbers $1, ..., n$, where n is the number of vertices of the graph G, such that

$$\sum_{\{u,v\} \in E} |F(u) - F(v)| \leq K?$$

3. Minimum cut of a graph [168]

Input data: a graph $G = (V, E)$ and a natural number K.
Question: does there exist a disjoint decomposition of V into two non-empty sets X and Y such that the number of edges $\{u, v\} \in E$ with $u \in X$ and $v \in Y$ is smaller than or equal to K?

4. Steiner tree [122]

Explanation of the problem. Let $\varepsilon > 0$. We are going to work with points in the euclidean plane: for two points x and y we use $d(x, y)$ to denote the euclidean distance of x and y rounded off to the nearest integer multiple of ε (see the end of the preceding paragraph).

If Z is a set of points, then by $T(Z)$ we denote a connected graph (Z, E) whose set of vertices is the set Z and for which the following sum:

$$D(Z) = \sum_{\{x,y\}\in E} d(x, y)$$

has the smallest value possible. (It is obvious that $T(Z)$ is a tree, which can be found by an easy application of the methods presented in § 6.2.)

It can happen that for two sets $X \subset Y$ we have $D(X) > D(Y)$ (try and find such an example!). Given a set X of points of the plane, then the set Y for which $X \subset Y$ and where $D(Y)$ has, under the considered condition, the smallest value possible, is called a *Steiner super-set* of X, and the tree $T(Y)$ is called the *Steiner tree* of the set X.

The following problem is NP-complete.

Input data: a set X of points in the plane with integer coordinates, a number $\varepsilon > 0$ and a natural number K.

Question: does there exist a set Y of points of the plane such that $X \subset Y$ and $D(Y) \leq K$?

5. One-processor scheduling [126]

Input data: a set T of tasks and for each task t three numbers: the running time $L(t)$ of processing of the task, the earliest time $I(t)$ of initialisation of the processing, and the latest time $F(t)$ of finishing the processing.

Question: is it possible to use one processor for all tasks, i.e. does there exist a mapping B from the set T to the set of all natural numbers (determining the beginning of the processing of the tasks) such that the following hold:

$$I(t) \leq B(t) \text{ and } B(t) + L(t) \leq F(t) \text{ for all } t \in T$$

and

$$\text{if } t_1 \neq t_2 \text{ and } B(t_1) \leq B(t_2), \text{ then } B(t_1) + L(t_1) \leq B(t_2)?$$

6. Scheduling with precedence [259]

Input data: a set T of tasks, for each task t the running time $L(t)$ of processing of t, a set P of processors, a partial ordering $\prec$ of the set T (where $t_1 \prec t_2$ means that t_2 cannot start being processed before the processing of t_1 has been finished), and a number F (the latest finishing time of all tasks).

Question: is it possible to process the given tasks using the given processors and respecting both the precedence and the time of finalisation? In other words, do there exist mappings

$$B: T \rightarrow \{1, ..., F\} \quad \text{and} \quad G: T \rightarrow P$$

with the following properties:

if $G(t_1) = G(t_2)$ and $B(t_1) \le B(t_2)$ with $t_1 \ne t_2$, then $B(t_1) + L(t_1) \le B(t_2)$,
if $t_1 \prec t_2$, then $B(t_1) + L(t_1) \le B(t_2)$,

and

$B(t) + L(t) \le F$ for each t?

Remark: this problem remains NP-complete even when restricted to the case of $L(t)$ equal to one for each t.

7. Quadratic congruence $[200]$

Input data: natural numbers a, b and c.
Question: does there exist a natural number $x < c$ with $x^c = a \bmod b$?

8. Quadratic diophantine equations $[200]$

Input data: natural numbers a, b and c.
Question: do there exist natural numbers x and y with $ax^2 + by = c^2$?

9. Graph thickness

Input data: a graph $G = (V, E)$ and a natural number k.
Problem: do there exist sets $E_1, ..., E_k$ such that $E_1 \cup ,..., \cup E_k = E$ and the graph (V, E_i) is planar for each $i = 1, ..., k$?

10. Chromatic index

Input data: a graph G.
Question: let D be the maximum degree of vertices of the graph G. Is it possible to colour the edges of the graph G by the colours $1, ..., D$ in such a way that two arbitrary incident edges have different colours? (It is obvious

that each such colouring requires at least D colours, and it has been proved that $D + 1$ colours are sufficient.)

The above list of NP-complete problems is not complete by a long way even after adding the problems listed in the exercises below. The interested reader is advised to consult the monograph [17] which presents an extensive list of NP-complete problems. Their range varies from practical problems such as the minimisation of the number of fictitious activities in the procedure used by the method PERT [182], or deciding whether a required computation can be performed by means of a given memory space [235], up to recreational puzzles such as the question whether a given set of words can form a crossword in a prescribed figure. Annotated references to the newest results on NP-completeness and some related areas can be found in the regular column of D S Johnson in the *Journal of Algorithms*.

Exercises

Prove the NP-completeness of each of the following problems.

1. **Covering by vertices**
   A graph $G = (V, E)$ and a natural number K are given. A decision is required as to whether there exists a set $X \subset V$ of K elements such that if $\{u, v\} \in E$, then either $u \in X$, or $v \in X$. (Use a reduction of the NP-completeness of the search of an independent set; see [168].)
2. **Dominant set of vertices**
   A graph $G = (V, E)$ and a natural number K are given. A decision is required as to whether there exists a set $X \subset V$ of K elements such that for each vertex $u \in V - X$ there exists a vertex $v \in X$ such that $\{u, v\} \in E$. (Use a reduction of the problem of the preceding exercise; see [168].)
3. A graph $G = (V, E)$ is given. Decide whether there exists a decomposition of E into two disjoint sets E_1 and E_2 such that none of the graphs (V, E_1) and (V, E_2) contains a triangle, i.e. three vertices pairwise connected by edges. (Prove this using theorem 7.4.7; see [168].)
4. A graph $G = (V, E)$ and a natural number K are given. Decide whether there exists a decomposition of V into K pairwise disjoint sets $V_1, ..., V_K$ such that for all $i = 1, ..., K$, each pair of distinct vertices of V_i is connected by an edge. (Prove this by reducing the problem of colouring by K colours; see [168].)
5. **Searching for a hamiltonian path**
   A graph G is given. Decide whether there is a hamiltonian path in G.

(Prove this by reducing the problem of searching for a hamiltonian circuit; see [168].)

6. **Open problem of travelling salesman**

   There are given a number K, a set M and a metric d on M. Decide whether there exists an ordering of the set M into a one-to-one sequence $x_1, ..., x_n$ such that $d(x_1, x_2) + ... + d(x_{n-1}, x_n) \leq K$. (Prove this by using the closed travelling salesman problem; see [168].)

7. **Minimum spanning tree with bounded degrees**

   A connected graph G with a positive edge-labelling C and numbers D and K are given. Decide whether there exists a connected subgraph T of the graph G such that the set of vertices of T is equal to that of G, the degrees of vertices in T do not exceed D, and the sum of labellings of the edges of T does not exceed K. (For $D = Z$, see exercise 5. The problem is easily solvable if the restriction of degrees is lifted.)

8. **The longest path in a graph**

   There are given a non-negatively edge-labelled graph G, vertices u and v of G, and a number K. Decide whether there is a path from u to v in G with the sum of labels of edges larger than or equal to K. (Use a reduction of exercise 5. The analogous problem concerning the shortest path is easily solvable, see §6.1.)

9. **The shortest path in a generally labelled graph**

   There are given a generally edge-labelled graph G, vertices u and v of G, and a number K. Decide whether there exists a path from u to v in G with the sum of labels of edges smaller than or equal to K. (Prove this by reducing the problem of the preceding exercise. For non-negative labels the problem is easily solvable, see §6.1.)

8

Classes of difficult problems

In this chapter we will first mention some of the properties of the class NP following from the assumption that $P \neq NP$, and we will study the computational complexity of an interesting problem: the determination of primality of natural numbers. Then we will investigate another class containing the class NP and defined on the basis of memory complexity. We are going to show that this class, denoted by $PSPACE$, is directly connected with the search for a winning strategy in a number of games, including some classical board games.

8.1 Properties of the class NP

In the present section we assume that $P \subsetneq NP$. In that case, theorem 7.4.10 says that the class P and the class NPC of all NP-complete problems are disjoint. We are faced with the question whether P and NPC form a covering of the class NP, as might seem to be the case after reading Chapter 7. However, it has been proved in [54] that for two arbitrary languages **J** and **K** such that $\mathbf{J} \lhd \mathbf{K}$ holds but $\mathbf{K} \lhd \mathbf{J}$ does not hold there exists a language **L** such that $\mathbf{J} \lhd \mathbf{L} \lhd \mathbf{K}$ holds, but neither $\mathbf{K} \lhd \mathbf{L}$, nor $\mathbf{L} \lhd \mathbf{J}$ hold. Assuming $P \neq NP$, we can find two such languages **J** and **K** in the class NP, and it then follows that the scale of complexities of problems belonging to the class NP is very rich, since by choosing $\mathbf{J} \in P$ and $\mathbf{K} \in NPC$, the complexity of the above language **L** lies somewhere between those of **J** and **K**, and then we can find a language between **J** and **L**, as well as one between **L** and **K**, and so on. The construction mentioned is based on the methods of the theory of recursive functions, and the resulting language **L** is nothing but a set of sequences of zeros and ones without any relation to 'everyday' problems, even if **J** and **K** both formalise such problems. No naturally described problem has, till the present time, been found in the class NP, which lies outside both of the classes P and NPC (at least assuming that $P \neq NP$). Some authors believe that the problem of determination of isomorphism of graphs could have that property and the same would hold for the problems of equal complexity, e.g. the problem of the determination of isomorphism of semigroups or automata.

Another class, denoted by *coNP*, is also often used. It is the class of all complements of the languages lying in *NP*.

Definition 8.1.1. A language **J** is an element of the class *coNP* if, and only if, the language $\{0, 1\}^* - \mathbf{J}$ belongs to the class *NP*. ●

The class *coNP* consists, essentially, of those problems for which it is simple to check the correctness of a negative solution. Let us first mention some properties of that class.

Theorem 8.1.2. $P \subset (NP \cap coNP)$.

Proof. If a language **J** is accepted by a deterministic acceptor M, then by interchanging the operations *ACCEPT* and *REJECT* in the program of the acceptor M we get an acceptor which accepts the complement $\{0, 1\}^* - \mathbf{J}$ of the language **J**. Thus, the complement lies in the class *P*, and consequently in the class *NP*, which proves that **J** belongs to the class *coNP*. It is now sufficient to use theorem 7.3.6. ●

Theorem 8.1.3. *If either* $coNP \subset NP$ *or* $NP \subset coNP$, *then* $NP = coNP$.

Proof. Assume that, for example, $coNP \in NP$, then for each language $\mathbf{L} \in NP$ the complement **K** of **L** lies in *coNP*, and hence in *NP*. It follows that $\mathbf{L} \in coNP$. This also shows that $NP \subset coNP$. The procedure when $NP \subset coNP$ is analogous. ●

Theorem 8.1.4. *Precisely one of the following assertions is true:*
(a) $P \subsetneqq NP \cap coNP$, $NPC \cap coNP = \emptyset$, $coNP \not\subset NP$, $NP \not\subset coNP$,
(b) $P = NP \cap coNP$, $NPC \cap coNP = \emptyset$, $coNP \not\subset NP$, $NP \not\subset coNP$,
(c) $P \subsetneqq NP = coNP$,
(d) $P = NP = coNP = NPC$.

The proof follows immediately from theorems 8.1.2 and 8.1.3. ●

The four possibilities of theorem 8.1.4 are depicted by figure 33. It is not yet known which of them actually takes place, but it is universally believed that the latter two do not. We have already mentioned the

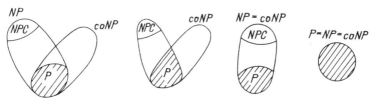

Figure 33.

problem '$P = NP$?'; the validity of the weaker equality $NP = coNP$ would mean that for all problems of the class NP, including the most difficult NP-complete ones, we would be able to prove quickly the correctness of negative solutions. At the present time, we are unable, for example in the case of a proof that a given graph does not contain a hamiltonian cycle, to proceed in any other way than by repeating the complete unsuccessful computation, which can demand a lot of time. We are going to speak below about possibilities (a) and (b), i.e. about the relationship of P and $NP \cap coNP$.

One of the problems which belong to the class $coNP$, and which do not require an unnatural formulation that would disguise the fact that a negative formulation of an NP-problem is actually being considered, is the problem of determining whether a given number n is a prime (although here, too, we have nothing more than a negation of the determination that the number is composite). A nice illustration of this problem is given in [220]: in order to disprove the hypothesis of Mersenne, 200 years old, that $2^{67} - 1$ is a prime, F Cole needed, in his own words, 'three years of Sundays'. However, when he lectured on his result at the meeting of the American Mathematical Society in 1903, all he needed to do was to write down the following equality

$$2^{67} - 1 = 193\ 707\ 721 \times 761\ 838\ 257\ 287.$$

An interesting feature of the problem of determination of primality is the fact that the problem itself also belongs to the class NP. In other words, it is not difficult to prove that n is a prime (once we know a proof, of course): this can be deduced from the following theorem of V Pratt [220] which is a rather simple consequence of the small Fermat theorem and the theorem on the primitive radix of a prime.

Theorem 8.1.5. *A natural number n is a prime if, and only if, there exist natural numbers p and $k_0, \ldots, k_m$ with the following properties:*
1. $k_0 = 1$ and $k_m = n - 1$,
2. *for each $i = 1, \ldots, m$ there exists a prime q_i such that $k_i = q_i k_{i-1}$, q_i is a divisor of $n - 1$ and $p^{(n-1)/q_i} \not\equiv 1 \bmod n$,*
3. $p^{n-1} \equiv 1 \bmod n$,
4. $m \leq \log n$. ●

In order to prove primality, it is thus sufficient to describe the sequence $p, k_0, \ldots, k_m$ and to verify the conditions 1, 2 and 3 of theorem 8.1.5. For the latter, the primality of the quotients k_i/k_{i-1} had to be proved, which is performed in the same way. Condition 4 says that the length of the 'proof'

$p, k_0, ..., k_m$ is smaller or equal to the size of the binary expression of the number n, or is larger by one at most. Due to this fact it can be shown that the verification of all the above conditions can be performed in running time bounded by a polynomial depending on the length of the number n written in binary form. As an example, we give a proof of primality of the numbers 251 and 47:

251: 1, 2, 10, 50, 250
 47: 1, 2, 46
 23: 1, 2, 22
 11: 1, 2, 10
  5: 1, 2, 4
  2: 1.

Other problems belonging to $NP \cap coNP$ are known: besides all problems of the class P this includes problems for which a minimax theorem holds, stating that for two functionals f and g which appear in the description of the problem we have $\min f = \max g$. As an example, let us mention theorem 6.3.12, or, more generally, the duality theorem of linear programming. If we want to prove that a number k can be the value of a functional f, then it is sufficient to describe x with $f(x) = k$; for the proof it is thus sufficient to perform a computation of the value of the functional. If, on the other hand, we want to prove that k cannot be a value of f, which usually means that $k < \min f$, then it is sufficient to describe y with $k < g(y) \leqq \max g = \min f$, and therefore, for the proof it is sufficient to perform a computation of the value of the functional g, and to rely on the relationship of f and g.

However, it has gradually turned out that all well-known problems with minimax theorems can be solved in polynomial-bounded time, as can be seen from the algorithms described in §6.3, and in the first place, from the algorithm solving the problem of linear programming in polynomial-bounded time; see [69]. The determination of primality is thus today the only important and well-known problem which we know belongs to the class $NP \cap coNP$ although we do not know whether it lies in P. Therefore, we cannot exclude the possibility that $P = NP \cap coNP$, particularly in view of the fact that there are indications that primality may also belong to the class P. We will discuss this in more detail in §10.6. It follows that whenever we manage to prove that a certain problem belongs both to NP and to $coNP$, it is worth while attempting to prove that the problem belongs, in fact, to the class P, or (which is simpler) trying to find such a proof in the literature.

Today a number of mathematicians, mostly among mathematical

logicians, study the relationship of the classes P, NP and $coNP$. The interested reader can consult the literature: $[61, 78, 143, 145, 164, 165, 173, 196, 197, 244]$.

8.2 Space complexity

In the present section we shall have a closer look at a class of problems defined by means of complexity of memory space.

Definition 8.2.1. *PSPACE* denotes the class of languages accepted in polynomial-bounded memory space; i.e. a language J belongs to *PSPACE* if, and only if, there exists a deterministic acceptor M accepting J and a polynomial p such that M requires memory space at most $p(n)$ for processing a word of length n. ●

We are interested not only in the properties of the class *PSPACE* but also in its relationship to the classes P and NP defined in the preceding chapter; the latter classes will now be denoted by *PTIME* and *NPTIME*, respectively. The basic relationship follows from theorem 2.2.4 which states that the requirement on memory space does not exceed that on running time. In other words we have the following.

Theorem 8.2.2. $PTIME \subset PSPACE$. ●

Conversely, knowing the requirements of the computation on memory space, we can obtain the following upper bound on the length of computation.

Theorem 8.2.3. *Suppose that a random access machine processes input data consisting of n numbers of absolute values not exceeding r, and that the computation requires memory space at most m. Then the computation either does not halt, or it halts after at most* $qn[2P(\max(n,r) + 1)]^m$ *steps, where q is the number of instructions of the program of the machine, and P is the polynomial of 2.2.3.*

Remark. For the present theorem as well as theorems 8.2.4 and 8.2.5 below, we assume the restriction of 2.2.3.

Proof. Due to 2.2.3, no memory register will, in the course of the computation, contain a number of absolute value larger than $\max(n, r)$. Owing to the bound on memory space, there are at most $[2P(\max(n,r) + 1)]^m$ states of the memory. Since the program register can only have q different values and it is possible to read n times from the input tape, it is evident that after performing more than $qn[2P(\max{(n, r)} + 1)]^m$ steps the machine would twice reach a completely identical configuration, and hence, it would reach an endless cycle. ●

In the same way as we introduced the class $NPTIME$ in the last chapter, we could define the class $NPSPACE$ of all languages accepted non-deterministically with a polynomial-bounded memory space. The following theorem due to W J Savitch [230] shows that this is not necessary.

Theorem 8.2.4. *A language* **J** *belongs to the class PSPACE if, and only if, there exists a non-deterministic acceptor which accepts* **J** *with a polynomial-bounded memory space.*

Proof. The inclusion $PSPACE \subset NPSPACE$ is obvious. Conversely, let M be a non-deterministic acceptor which accepts **J** and which, for a given polynomial p, processes a word of length n using memory space at most $p(n)$. In order to be able to find out whether M accepts a given word w or not, it is sufficient, due to theorem 8.2.3, to answer the question whether M can change its initial configuration determined by the word w to an accepting configuration which performs the instruction $ACCEPT$ after at most qnN^m steps, where $N = 2P(\max{(n, r)} + 1)$ and q and r are constants given by the acceptor M.

We are going to show how one can construct an algorithm A which finds out deterministically whether, for a given number k and given configurations C and C' of the machine M, the machine can change configurations from C to C' after at most 2^k steps and using memory space at most m. For $k = 0$, it is sufficient to verify the conditions introduced in § 2.1, which is lengthy but easy. For larger k, we shall proceed recursively: we shall define an auxiliary memory region capable of keeping the description of the next configuration D, and then we will search successively all possible values of D. The fact that the memory space is bounded by the number m will guarantee that there are finitely many values of D only. For each possible value of D we then find out whether the machine can change the configuration C to D and the configuration D to C' both in at most 2^{k-1} steps. This is performed by a double recursive application of the algorithm A.

For the computation we thus have to describe k auxiliary configurations corresponding to the k levels of calling the algorithm A. In our case, it is necessary that $2^k \geq qnN^m$, and for this it is sufficient to choose

$$k = \lceil \log qnN^m \rceil \leq \lceil m \log qnN \rceil.$$

Recalling that $m = p(n)$, we see that there exists a polynomial bound on k in dependence on n. Since the memory space necessary for the description of one configuration is, due to the bound $m = p(n)$ on the memory space, also a polynomial function of n, we conclude that the algorithm A, however slow it may be, requires a polynomial-bounded memory space too, and the computation according to it is deterministic. ●

The equality $PSPACE = NPSPACE$ is one of the few non-trivial equalities which we are able to prove for the relationship of the classes of problems defined according to bounds on the running time or memory space.

The above theorem makes it possible to prove the following relation of the classes under consideration.

Theorem 8.2.5. $PTIME \subset NPTIME \subset PSPACE$.

Proof. The left-hand inclusion has already been proved in 7.3.6 (where, instead of $PTIME$ and $NPTIME$, we used the symbols P and NP). Theorem 2.2.4 remains valid, including the proof also in the non-deterministic case. It follows immediately that $NPTIME \subset NPSPACE$ which, combined with the equality $NPSPACE = PSPACE$ of theorem 8.2.4, yields the right-hand inclusion above. ●

As usual, we are not yet able to decide whether $NPTIME = NPSPACE$, but the equality is considered not to be provable.

8.3 *PSPACE*-completeness

A study of the class $PSPACE$ can be performed using the concept of reduction $\lhd$ as introduced in definition 7.4.1 in the same manner as the class $NPTIME$ (or NP) was investigated in Chapter 7. A number of analogous results can be proved. First of all, we will show that there also exist problems of maximum complexity in $PSPACE$, which means the following.

Definition 8.3.1. A language $\mathbf{J}$ is said to be *PSPACE-hard* if for each language $\mathbf{K} \in PSPACE$ we have $\mathbf{K} \lhd \mathbf{J}$. A *PSPACE*-hard language $\mathbf{J}$ such that $\mathbf{J} \in PSPACE$ is said to be *PSPACE-complete*. ●

The following simple assertions can be proved.

Theorem 8.3.2. *Precisely one of the following two possibilities occurs*:

(a) $NPTIME = PSPACE$ and every NP-complete problem is PSPACE-complete.

(b) $NPTIME \neq NSPACE$ and no NP-complete problem is PSPACE-complete.

Proof. If a language $\mathbf{J} \in NPTIME$ is PSPACE-complete, then for each language $\mathbf{K} \in PSPACE$ we have $\mathbf{K} \triangleleft \mathbf{J}$, and hence, $\mathbf{K} \in NPTIME$. It is now sufficient to use theorem 8.2.5. ●

Theorem 8.3.3. *If a problem* $\mathbf{J}$ *is PSPACE-complete, then each problem* $\mathbf{K}$ *with* $\mathbf{J} \triangleleft \mathbf{K}$ *is PSPACE-hard.*

Proof follows immediately from the transitivity of $\triangleleft$. ●

An important theorem which is an analogous result to Cook's theorem 7.4.6 has been proved by L J Stockmeyer and A R Meyer [246].

Theorem 8.3.4. *The problem of determining whether a quantified formula of the following type*

$$\exists x_1 \forall x_2 \ldots \exists x_{2n-1} \forall x_{2n} f(x_1, \ldots, x_{2n})$$

in conjunctive normal form is true is PSPACE-complete.

Proof. We will first show that the problem belongs to the class *PSPACE*. Let $a_1, \ldots, a_i$ for $i \leq 2n$ be logical values true and false. Then we say that f is satisfied with respect to $(a_1, \ldots, a_i)$ if one of the following cases takes place:

either $i = 2n$ and $f(a_1, \ldots, a_{2n}) = $ true,
or $i < 2n$ is even (and hence x_i is quantified by $\forall$) and f is satisfied both with respect to $(a_1, \ldots, a_i, $ true) and $(a_1, \ldots, a_i, $ false),
or $i < 2n$ is odd (and hence x_i is quantified by $\exists$) and f is satisfied either with respect to $(a_1, \ldots, a_i, $ true), or to $(a_1, \ldots, a_i, $ false).

It is easy to verify that a formula of the type considered is true whenever f is satisfied with respect to the empty sequence of truth values. The above procedure yields a recursive algorithm which needs, in order to find out whether the formula f holds, nothing more than a description of the formula f, the variables $a_1, \ldots, a_n$, and several auxiliary variables. Consequently, the computation can be performed in a linearly bounded memory space.

Next, we shall verify that the above problem is *PSPACE-hard*. Assume that a non-deterministic acceptor is given, working in memory space bounded by a polynomial p, and let w be a word.

Denote by n the length of the word w and by m the value $p(n)$. For a description of one configuration of the machine M we only need logical variables, the number of which is bounded by a polynomial function of m, and hence, of n; it is sufficient to use the variables of the proof of theorem 7.4.6 and to omit the time index t. It can also be seen from the proof of that theorem how one can construct formulae h_1 and h_2 the variables of which represent a system capable of describing configurations of the machine M and such that $h_1(C)$ and $h_2(C)$ is true if, and only if, the system C describes the initial configuration determined by the word w, or an accepting configuration performing the instruction *ACCEPT*, respectively. Analogously, one can construct, as in the proof of theorem 7.4.6, a formula q_0 depending on two systems C and C' of variables capable of describing configurations in such a way that $g_0(C, C')$ is satisfied if, and only if, C and C' describe two configurations such that the machine M can proceed from the first one to the latter one within one step of the computation.

We will now show how to construct a formula which, for descriptions C and C' of configurations, is satisfied if, and only if, the machine M can change the first configuration to the latter one in at most 2^k steps. That quantified formula will be denoted by $g_k(C, C')$ and will be constructed (using auxiliary variables D, E and E') as follows:

$$\exists D \; \forall E \; \forall E' \big[\big(\big(C = E \wedge D = E'\big) \vee \big(D = E \wedge C' = E'\big)\big) \Rightarrow g_{k-1}(E, E')\big] \, .$$

In fact, this formula is satisfied if, and only if, M can change both the configuration C to D in at most 2^{k-1} steps, which is expressed by

$$\big(C = E \wedge D = E'\big) \Rightarrow g_{k-1}(E, E') \, ,$$

and the configuration from D to C' in at most 2^{k-1} steps, which is expressed by

$$\big(D = E \wedge C' = E'\big) \Rightarrow g_{k-1}(E, E') \, .$$

This, together with the existential quantification of D, yields the required property (where the equality $C = E$ represents the requirement of correspondence of variables in the systems C and E, the quantification $\exists D$ means that all variables of the system D are quantified by $\exists$, etc).

If in a non-deterministic computation one configuration appears twice, it does not necessarily mean that we get into an endless cycle, since we can leave such a cycle by a non-deterministic jump instruction. However, the computation can be shortened by deleting the whole cycle which starts and terminates in the same configuration. This will result in a shorter admissible computation in which the given configuration is not repeated. Consequently, if the machine M accepts the word w, it can be proved analogously to the proof of theorem 8.2.3 that there exists an admissible computation determined by the word w and terminating by a performance of the instruction *ACCEPT* after at most qnN^m steps, where $N = 2P(\max(n, r) + 1)$. It is now clear that the following formula:

$$\exists C \, \exists C'(h_1(C) \wedge h_2(C') \wedge g_k(C, C')), \quad \text{where} \ k = [m \log(qnN)],$$

is true if, and only if, the machine M accepts w. Analogously to the proof of theorem 8.2.4 it can be shown that the whole construction can be performed in polynomial-bounded time. The resulting formula is not in the required normal form, but it can be transformed (in polynomial-bounded time again) to the normal form — we refrain from performing this since the formula is very complicated. ●

There is an interesting interpretation of the determination of validity of the quantified formula

$$(1) \qquad \exists x_1 \, \forall x_2 \ldots \exists x_{2n-1} \, \forall x_{2n} f(x_1, \ldots, x_{2n}).$$

Let two players, say black and white, alternately choose the truth value of the variables $x_1, \ldots, x_{2n}$. The white player begins, and hence he chooses the values of the variables quantified by $\exists$, whereas the black one chooses the values of the variables with the quantifier $\forall$. If at the end of the choice we have $f(x_1, \ldots, x_{2n}) = \text{true}$, the white player wins, otherwise the black one wins.

It is clear that the quantified formula (1) is true if, and only if, the white player wins provided he has played well, regardless of the moves of the black player. Conversely, the black player has a winning strategy when the formula (1) is not true. Thus, theorem 8.3.3 can be formulated as the *PSPACE*-completeness of the problem of searching for an optimum strategy of one of the players of a certain game.

It turns out that problems of the above type are typical representatives of *PSPACE*-complete problems. We will now show an analogous result for a game which is not very interesting in its own right, but the performance of which allows us to prove *PSPACE*-completeness of solving game situations in some classical board games such as draughts and go. The description of the game is as follows: the 'board' is a directed graph G in which an initial vertex v_0 has been chosen. At the beginning of the game a piece is placed at v_0 which, in the course of the game, can be moved onto other vertices of the graph. A move of the game consists of moving the piece from one vertex to another subject to the following rules:

1. the piece can be moved from x to y only if (x, y) is a directed edge of the graph G,

2. the piece must not be moved to a vertex where it has already been in the course of game.

The game is played by two players, white and black. The player with no possibility of a move according to the rules of the game loses.

Rule 2 guarantees that the number of moves is smaller than the number of vertices of the graph, and hence, that the game is finite. Consequently, one of the players has a winning strategy, i.e. if he plays in the optimum way then he must win regardless of the strategy of his opponent. It is our task to prove the following theorem [232].

Theorem 8.3.5. *The decision as to which player has a winning strategy in the game described above for a given directed graph* **G** *and a given vertex* v_0 *of* **G** *is a PSPACE-complete problem.*

Proof. The fact that the problem under consideration belongs to the class *PSPACE* can be proved by an analogous procedure to that of the proof of theorem 8.3.4, and we thus leave that part of the proof to the reader as an exercise.

The more important part of the theorem is the fact that the above problem is *PSPACE*-difficult. Suppose we are given a formula

(2) $$\exists x_1 \forall x_2 \dots \exists x_{2n-1} \forall x_{2n} f(x_1, \dots, x_{2n}),$$

for $f(x_1, \dots, x_{2n}) = (a_{11} \lor \dots \lor a_{1m_1}) \land \dots \land (a_{k1} \lor \dots \lor$

where the variables a_{ij} are either of the form x_p or $\neg x_p$ for suitable p. We are going to construct a graph **G** and a vertex v_0 of **G** in such a way that

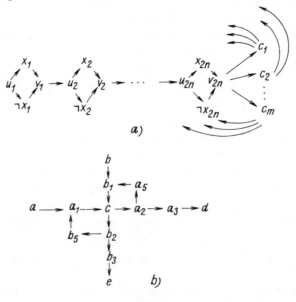

Figure 34.

formula (2) is true whenever in the game determined by G and v_0 the white player has a winning strategy.

The basis of the graph is formed by vertices which correspond to the variables x_i and their negations $\neg x_i$ and furthermore vertices by u_i and v_i for $i = 1, \ldots, 2n$ connected by directed edges in the manner indicated in figure 34a. We choose the vertex u_1 as v_0. Let us have a look at the start of the game. The white player chooses either x_1 or $\neg x_1$ and moves to the corresponding vertex, then the black one moves to v_1 and the white one to u_2. Afterwards, the black player chooses either x_2 or $\neg x_2$ and moves to the corresponding vertex, the white one moves to v_2, the black one to v_3, and then the white one can choose between x_3 and $\neg x_3$, and so on. The procedure is thus the same as that of the logical game above according to formula (2). It is the white player who enters v_{2n}, and hence, the black one chooses among the c_i and he moves the piece there. This corresponds to the choice of the ith clause $(a_{i1} \vee \ldots \wedge a_{ik})$ in formula (2) above, and hence, if the white player wants to win, he must have the possibility of performing, in the next continuation, an activity which is possible only if all factors of formula (2) are true. The edges with initial vertex c_i in the graph of figure 34a are interconnected in such a way that for all $j = 1, \ldots, k_i$ there is a directed edge (c_i, b_{ij}) in the graph G, where $b_{ij} = \neg x_p$ if $a_{ij} = x_p$, and $b_{ij} = x_p$ if $a_{ij} = \neg x_p$.

If the transition through the basic part of the graph and the corresponding values of the logical formulae are such that for each i there exists a j with $a_{ij} =$ true, and hence $(a_{i1} \vee \ldots \vee a_{ik_i})$ is true, then regardless of the choice of c_i by the black player, the white one can continue to a b_{ij} in which the piece has not yet occurred. By then, following the rules of the game, the black player is forced to move to a certain vertex v_p in which the piece has already been, in other words, he loses. Conversely, if for some i all values of a_{ij} are equal to false, and thus the formula above is not satisfied, the optimum choice of the black player is the corresponding vertex c_i since this does not leave any correct move open to the white player. ●

The above game can be simplified, while preserving the complexity of determining which player has a winning strategy.

Theorem 8.3.6. *The assertion of theorem 8.3.5 remains valid even when we restrict ourselves to planar graphs.*

Proof. We draw the graph constructed in the proof of the preceding theorem in the plane. If the graph is not planar, there necessarily occur crossings of the edges with the initial vertices c_i. The crossing of two edges, say, (a, d) and (b, e), will be substituted by the graph of figure 34b. Suppose one of the players, whom we denote by X, is to move, and he wants to continue to d. The game then has the following continuation in the graph of figure 34b: X moves to a_i, the other player Y moves to c, and should X not move to b_2, then Y would move to b_5 and X would lose — thus, X has to move from c to a_2. If Y now were to move from a_2 to a_5, then X would move to b_1, and Y would lose. Thus, Y has to move to a_3, X continues to d, and we see that the game proceeds in the same way as in the original graph. The symmetry of the auxiliary graph makes it unnecessary to deal with the case of the edge (b, e). ●

Since the graphs considered in the preceding theorems are to be arbitrarily large, the usual board games cannot be used as a model of a play of the game discussed above, because the common games, e.g. chess, are bounded to a certain size of the board. However, for some games, e.g. draughts and go, which are played with one sort of pieces, the size of the board is a matter of convention only (and, in fact, beginners at go often play on smaller boards). If we admit an unbounded size of the board, it can be shown that the planar games of theorem 8.3.6 have a model in game situations both in draughts [107] and in go [189].

The value of such results is limited since, for example, a player of go will have difficulty in understanding why the complexity of his game requires arbitrarily large boards when the game is already so rich on the classical 19 by 19 board. However, the above results are interesting by showing the combinatorial richness of classical board games.

A majority of the well-known *NPSPACE*-complete problems require the determination of whether it is possible to reach a certain goal in spite of the existence of an opponent who tries to prevent it. This distinguishes the *NPSPACE*-complete problems from the problems which we can prove belong to the class *NPTIME*.

An exception to the above rule is the pebbling problem of graph theory which is *NPSPACE*-complete although no explicit opponent is given; see [131 and 192]. The problem is described as follows.

There are given a directed graph G with a base vertex v and k pebbles which can be placed on the vertices of the graph G or deleted from the vertices subject to the following rules.

1. A pebble can be placed on a vertex only if all predecessors of the vertex are occupied by pebbles.

2. A pebble placed on a vertex can be deleted at any instant.

The task is to determine whether the given number of pebbles is sufficient for the base vertex v to be covered by a pebble after a certain number of steps 1 and 2.

This problem is a model of the distribution of the memory space of a computer in the course of computation. The pebbles represent the memory registers of the computer, the vertices correspond to the partial results obtained in the course of computation, and an edge $x \rightarrow y$ means that the value of x is needed for the determination of the value of y. Placing a pebble on a vertex means storing a partial result in the corresponding position of the memory. The number of pebbles means the size of the memory space assigned to the computation, and the number of steps 1 and 2 above corresponds to the length of the computation.

The pebbling problem is also useful in providing a simple instrument for the investigation of the general relationship of the memory space requirements and the running time requirements of computation. Some results suggest, see [216], that by decreasing the number of pebbles by one the goal can still be reached, but the number of necessary steps can increase astronomically.

9

Heuristic methods

In the preceding two chapters we have explained why for a number of difficult problems no algorithms are known which would find the required solution in polynomial-bounded time. In computer practice, the solution of those problems is performed by means of heuristic methods designed in such a way that a suitable solution is found as often as possible. The primary criterion of quality of those algorithms is how good are the actual applications of the methods in practice. Heuristic procedures are usually not compared with the algorithms for searching for the optimum solution, but rather they are compared among themselves with the aim of finding the best one. Sometimes several heuristic methods, based on different principles, are also used in one solution of a problem. Furthermore, the number of theoretical results dealing with a probabilistic analysis of heuristic algorithms is steadily increasing.

The aim of this chapter is to describe the current methods of solution of four important problems: determination of the chromatic number of a graph, searching for a hamiltonian circuit, the travelling salesman problem, and the problem of isomorphism of graphs.

9.1 Graph colouring

When dealing with problems which directly or indirectly include the task of colouring the vertices of a graph by the least number of colours (e.g. construction of time schedules) we often encounter input data of large extent, and the problems of that kind belong to the most difficult ones of practical computation. That is, they are NP-complete, and fast approximate algorithms which would be sufficiently precise are not known for them: for example, an algorithm which would colour a graph in such a way that the number of colours used would never exceed the chromatic number of the graph by more than 33 per cent would solve the problem of theorem 7.5.1 which is NP-complete.

The most widely used type of algorithm for an approximate determination of the chromatic number is described by the following scheme.

Algorithm 9.1.1. Sequential colouring of a graph

Input data: a graph G.

Task: colour G by the least possible number of colours.

Auxiliary variable: K, the number of colours already used.

1. [*Initiation*] $K := 0$.

2. [*Cycle*] While there exists a non-coloured vertex of the graph, repeat operations 2a to 2c.

2a. [*Choose a vertex*] Choose a vertex v_0 of the graph G which has not yet been coloured.

2b. [*Determine the colour*] Find the minimum natural number b such that no vertex connected by an edge with v_0 has already been coloured by the colour b.

2c. [*Colour the vertex v_0*] Colour v_0 by the colour b, and if $K < b$, put $K := b$. ●

There exist several basic ways of choosing the vertex v_0 in step 2a above.

(i) A random choice of the vertex v_0. Usually, the vertices are ordered at the beginning of computation in a random way, and the vertex v_0 is chosen as the first non-coloured vertex in that ordering.

(ii) The vertices are ordered at the beginning of computation in such a way that their degrees form a non-increasing sequence, and v_0 is chosen to be the first non-coloured vertex in that ordering. It means that we first colour the vertices of large degrees, leaving the vertices with small degrees to the end of the colouring. See the paper by D J Welsh and M B Powell [266].

(iii) This is analogous to (ii) but the initial ordering is performed in a different way: the ordering $v_1, ..., v_n$ of vertices is determined backwards, v_n is a vertex of the smallest degree, and given $v_{j+1}, ..., v_n$, then v_j is a vertex connected with the least number of vertices whose position in the sequence has not yet been determined. The vertices with large degrees have, therefore, a higher priority again. See the paper by D W Matula *et al* [204].

(iv) For each vertex v denote by $D(v)$ the number of colours which have already been used for the colouring of the neighbours of v. Choose as v_0 an uncoloured vertex with the largest possible value of $D(v)$. When there is a larger number of such vertices, we choose that which has the largest number of uncoloured neighbours. See the paper by D Brélaz [66].

The above methods of choosing v_0 are mostly based on the principle that one should first colour those vertices for which the task is the most difficult. Other methods of making the choice are also known, for example, choosing in an order given by some methods of search (suitable for bipartite graphs) or by making use of degrees of higher order (see the paper by M R Williams [271]), etc.

A certain improvement of algorithm 9.1.1 can be achieved by means of a trick which is well known from some proofs of theorems of graph theory. Suppose that we are colouring the graph of figure 35 sequentially from left

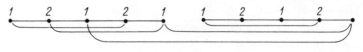

Figure 35.

to right. Before colouring the last vertex, the depicted situation takes place: even if the graph is bipartite, we shall be forced to use a third colour. By analysing the situation, we see that the subgraph formed by all already coloured vertices has two components. In any of them we can interchange the colours 1 and 2 without obtaining a contradiction of the definition of graph colouring. If we interchange the colours, say, in the left-hand component (consisting of the first five vertices from the left), it is possible to use colour 1 to colour the last vertex.

In general, we proceed as follows.

Algorithm 9.1.2. Sequential colouring of a graph with colour interchange
Proceed as in 9.1.1 except that step 2c is performed as follows:

2c'. [*Interchange colours*] (Let b be the colour by which the vertex v_0 is to be coloured, and let K be the number of colours used so far.)

If $b \leq K$, we colour v_0 by the colour b; else, we perform the following operations A, B and C for each pair i and j of numbers with $1 \leq i < j \leq K$:

A: denote by V_{ij} the set of all vertices coloured by either i or j;

B: construct the components of the full subgraph of the graph G determined by the set of vertices V_{ij};

C: if in each of the components all neighbour vertices of the vertex v_0 are coloured by only one of the colours i and j, then in all those components in which the neighbours of v_0 are coloured by (say) the colour j we interchange the colours i and j. Then we colour the vertex v_0 by the colour j and we terminate the performance of step 2c'.

If in the course of performing A, B and C we have not managed to perform the interchange of colours and the colouring of v_0, then we colour v_0 by the colour b, and we put $K := b$. ●

The choice of the vertex v_0 can be performed in the same way as in algorithm 9.1.1.

The results of a detailed investigation of the graph colouring algorithms given in [66] show that the ratio of the result to the optimum number of colours given by the chromatic number which was determined by a precise method is as presented by the following table (for random graphs of less than 100 vertices):

9.1.1(ii)	9.1.1(iii)	9.1.1(iv)	9.1.2(iii)	9.1.2(iv)
1.12–1.16	1.095–1.13	1.036–1.069	1.044–1.077	1.023–1.055

Although these results are very optimistic (precision better than 5.5 per cent in the last algorithm) we have to recall that the table is related only to the type of random graphs used, corresponding to the construction we describe in § 10.4 below, and the results can be quite different in other cases (see § 10.5).

Whereas algorithms 9.1.1 and 9.1.2 considered all vertices and assigned colours to them, the following procedure assigns, conversely, vertices to the colours. It is based on the fact that all vertices coloured by one colour form an independent set.

Algorithm 9.1.3. Colouring of a graph by means of independent sets
Input data: a graph G.
Task: colour the vertices of G by the least possible number of colours.
Auxiliary variables: X, an independent set of the graph G, and K, the number of colours used.
  1. [*Initiation*] $K := 0$.
  2. [*Cycle*] While at least one vertex is left in the graph G, perform steps 2a to 2c.
  2a. [*Choose a new colour*] $K := K + 1$.
  2b. [*Determine an independent set*] We determine an independent set X of the graph G, as large as possible.
  2c. [*Colouring of vertices*] Colour all vertices of the set X by the colour K and delete all those vertices and all edges incident with them from the graph G. ●

In algorithm 9.1.3, we must specify the way in which a large independent set of vertices is to be determined; since this problem is important in its own right, we formulate it as a separate algorithm.

Algorithm 9.1.4. Searching for an independent set
Input data: a graph G.
Task: find an independent set of the graph G as large as possible.
Auxiliary variable: Y, a set of vertices of the graph G.

1. [*Initiate*] $X := \emptyset$ and $Y := V(G)$.

2. [*Cycle*] While $Y \neq \emptyset$ perform the following operations.

2a. [*Choose the next element of* X] Choose $v_0 \in Y$ and put $X := X \cup \{v_0\}$, delete v_0 and all neighbour vertices of v_0 from the set Y. ●

Algorithm 9.1.4 also has several variants according to the manner in which the vertex v_0 is chosen:

(i) we make a random choice of v_0, or choose the first possible element of a random ordering of vertices determined before computation;

(ii) before starting the computation we order the vertices according to degree from the smallest to the largest, and we always choose as v_0 the first possible element in that ordering;

(iii) we choose as v_0 a vertex for which the instant value of $\deg_G(v, Y)$ is minimum.

There are considerable differences in the speed of the individual variants above, but this is sometimes balanced by a substantial difference of the quality of the solution. (See, for example, [184].)

Another colouring method is sometimes used, based on that described by A A Zykov [279].

Algorithm 9.1.5. Colouring of graphs by merging vertices

Input data: a graph G.

Task: find a colouring of the graph G by the least possible number of colours.

1. [*Cycle*] Until G is reduced to a complete graph (in which each pair of distinct vertices is connected by an edge) perform steps 1a and 1b.

1a. [*Choose vertices to be merged*] Choose vertices v_1 and v_2 of the graph G which are distinct and are not connected by an edge.

1b. [*Merge the vertices*] Substitute the vertices v_1 and v_2 by a new vertex v_0 which is connected to all vertices of the graph G to which either v_1 or v_2 were connected.

2. [*Colour*] Each vertex of the resulting graph is coloured by a different colour. The vertices of the original graph are coloured according to the colours of the vertices through which they were eventually substituted.

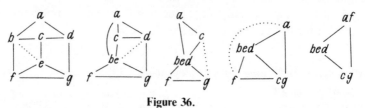

Figure 36.

Figure 36 illustrates this simple method. The vertices a and f will be coloured by one colour, the vertices b, e and d by another one, and the vertices c and g by a third one. (The dotted lines do not denote edges but indicate the vertices to be merged.)

For choosing the vertices v_1 and v_2 in step 1a we can again use various numerical characteristics, e.g. the degrees of vertices, etc.

We finally give a way of searching for the precise value of the chromatic number of graphs [67]. This procedure is a sample of the methods described in § 4.3 in the purest form, not disguised by using further improvements and tricks. We will decide on a premature termination of a branch of the computation only when otherwise the rules given by the definition of colouring would force us to use a higher number of colours than that given by an upper bound U of the chromatic number.

Algorithm 9.1.6. Optimum colouring of graphs
Input data: a graph G of n vertices.
Task: colour G by the minimum number of colours.
Auxiliary variables:

U, an upper bound on the chromatic number of the graph,

L, a lower bound on the chromatic number of G,

B_k for $k = 1, ..., n$, a set of colours (i.e. numbers $1, ..., n$) which are considered for colouring the kth vertex,

K, the number of colours already used,

K_k for $k = 1, ..., n$, the value of K at the instant of colouring the kth vertex.

1. [*Determine the bounds*] Determine an upper bound U and a lower bound L of the chromatic number of G.

2. [*Order the vertices*] Order the vertices of the graph G into a sequence $v_1, ..., v_n$.

3. [*Initiation*] $K_1 := 1$, $K := 1$, $k := 2$; colour v_1 by the colour 1; $B_1 := \emptyset$.

4. [*Determine B_k*] The set B_k is obtained from the set of all numbers $1, ..., m$, where $m = \min(K + 1, U - 1)$, by deleting all colours already used for the colouring of those vertices v_i for $i < k$ which are connected with v_k by an edge. (There are three reasons for the choice of that set:

— if none of the colours $1, ..., K$ can be used for colouring the vertex v_k, then it is sufficient to use the colour $K + 1$;

— if we want to obtain a better colouring than the one already known using U colours, we cannot use the colours $U, U + 1, ..., n$;

— the vertex s_k cannot be coloured by the same colour as any of its neighbours.)

5. [*Test of termination of a branch of computation*] If $B_k = \emptyset$, go to step 9.

6. [*Colour the vertex v_k*] Determine the minimum number which is a member of U_k, colour the vertex v_k by the colour i, put $B_k = B_k - \{i\}$, and if $K < i$ put $K := i$ (a new colour has been used), and put $K_k := K$.

7. [*Has an improvement been found?*] If $k < n$, put $k := k + 1$ and go to step 4, else store the constructed colouring (which is the best of the colourings found so far) and put $U := K$. If we get $L = U$, then U is equal to the chromatic number of the graph, and the computation is terminated; else, continue to step 8.

8. [*Determine the beginning of a new branch of computation*] Determine the smallest j such that the vertex v_j is coloured by the colour U. Put $k := j - 1$ and go to step 5. (If we want to achieve an improvement of the colouring by U colours, then the colour of the vertex v_j has to be changed, which, however, is impossible without changing the colour of some of the vertices v_k for $k < j$. We choose $k := j - 1$.)

9. [*Backwards procedure in the solution tree*] If $k = 1$, terminate the computation: U is equal to the chromatic number of the graph. Else, put $k := k - 1$, $K := K_k$, and go back to step 5. ●

The initial ordering for step 2 can be chosen in various ways — the methods (ii) and (iii) presented after algorithm 9.1.1 above turn out to be suitable. As the bounds in step 1 we can choose $U = n$, corresponding to the trivial colouring of the vertex v_i by the colour i, and $L = 2$, assuming that G has at least one edge (otherwise there is nothing to compute). If we use the upper bound $U = n$, then algorithm 9.1.6 works first as the sequential procedure 9.1.1 based on the given initial ordering of the vertices and it thus creates a good initial approximation U for the next branch. Since the length of the computation of algorithm 9.1.6 can be substantially reduced by using a good upper bound U, it is of some advantage to apply at the beginning of computation a better heuristic for the determination of U than that on which the computation of 9.1.6 is based (for example, 9.1.2 and 9.1.3 (iii)). The lower bound has a weaker influence on the length of computation than the upper bound. If we determine it as the size of a clique of the graph G found by some heuristic procedure then, assuming that $\gamma(G) < \chi(G)$ (which is often the case, particularly for large graphs), the lower bound does not influence the length of computation at all.

9.2 Hamiltonian path and circuit

In this section we give some of the methods used for searching for a hamiltonian path or circuit in a graph. Each procedure of §9.3 below, devoted to the travelling salesman problem, can be applied too, of course, since the travelling salesman problem is more general than the present one, but in this case, the coefficients of the matrix of distances would have two possible values only, which often leads to a collapse of the algorithm to a trivial one whose heuristic power is not very effective.

The basic type of algorithm for searching for a hamiltonian path or circuit is based on the methods described in § 4.3. By an *admissible partial solution* is here meant a path $x_0, ..., x_k$ in the graph under consideration, which we are trying to prolong in order to find a hamiltonian circuit or path. We prolong it by successive addition of edges leading from the end-vertex of the path to vertices which have not yet been used. If no such prolongation exists, we shorten the path to $x_0, ..., x_k$, and we try to find a prolongation in a different way. The computation terminates at the instant when either the required solution has been found, or all the possibilities of prolongation have been attempted unsuccessfully.

It is usually necessary to apply various improvements, the goal of which is to reduce the often too lengthy computation. A detailed description of such improvements can be found in [10]. The basic idea is to use two sets J and N of edges which contain, at the instant of processing a partial solution $x_0, ..., x_k$, all edges which lie or cannot lie, respectively, on a prolongation of $x_0, ..., x_k$ into a hamiltonian path or circuit. When prolonging the given partial solution, one proceeds as follows: if there exists an edge in J with one end-vertex x_k and whose other end-vertex has not yet been used, then we are obliged to use that edge for a prolongation. If no such edge exists, it is possible to choose a prolongation only among edges not lying in N. With the growth of J and N the number of possibilities which have to be considered decreases, and this speeds up the computation. With each prolongation of the partial solution, the sets J and N grow according to the following rules which are iteratively repeated when needed:

(a) the edges $\{x_0, x_1\}, ..., \{x_{k-1}, x_k\}$ are elements of J; if $k < n$ (where n is the number of vertices of the graph), then $\{x_0, x_k\}$ is an element of N provided that it is an edge of the graph,

(b) if two edges with an end-vertex x belong to J, then all the remaining edges with that end-vertex are inserted into N,

(c) if a vertex x is incident on exactly two edges which are not elements of N, then the two edges must be elements of J,

(d) if some vertex x is incident on a unique edge which is not an element of N, or conversely, if x is incident with at least three vertices which are elements of J, then the path $x_0, ..., x_n$ cannot be prolonged to a hamiltonian path or circuit, and we must shorten it.

When shortening the path, it is necessary to restore the sets J and N to the form which they had before the prolongation by the edge $\{x_{k-1}, x\}$. It is thus reasonable to store, for each edge lying in J or N, the information specifying in which stage of the prolongation the edge had been inserted into J or N. If the sets J and N are implemented as lists ordered by the number of stages of storing the edges, then a reconstruction of the original form of the above sets is performed by deleting the ends of those lists, which is a simple and rapid operation.

The whole method above just describes the usual way of 'manual' search for a hamiltonian path or circuit. It speeds up the computation considerably in cases when the graph does not have too many edges since then rule (c) is often applied. For dense graphs the improvement is less outstanding.

Results of Chapter 7 show that one cannot expect that the algorithm, the idea of which we have just described, or any other procedure which finds a hamiltonian path or circuit whenever they exist, will compute quickly enough to make it possible to process large graphs. For example, an empirical formula for the unimproved variant of prolongation just described is given in [10]: it follows from experiments on the CDC 6000 computer that 1 second is sufficient to process a graph of approximately 25 vertices, and each extension of the set of vertices by 2 more then doubles the running time. The procedure is therefore not suitable for processing large graphs. Consequently, quick heuristics are used which make it possible to process graphs of hundreds of vertices, but for which there is a non-zero probability that no result is found although the hamiltonian path or circuit being searched for does exist in the graph. When diminishing the probability of an unsuccessful search we usually lose computation speed, and thus when designing or choosing a procedure, we have to find a compromise depending on the actual situation.

We will now describe a good heuristic used by L Pósa in his paper [219].

Algorithm 9.2.1. Searching for a hamiltonian path
Input data: a graph G.
Task: find a hamiltonian path or circuit in G.
Auxiliary variables:
P, a segment of the path or circuit,
S, a set of vertices of the graph G.

1. [*Initiate*] Choose an arbitrary vertex v_0 of the graph. Denote by P the path formed by the single vertex v_0, and put $S := \emptyset$.

2. [*Prolong*] If the path P is equal to the sequence $v_0, \ldots, v_k$ and if there exists a vertex v distinct from the vertices lying in P and such that $\{v_k, v\}$ is an edge of the graph G, then prolong P to the path $v_0, \ldots, v_k, v$; and put $S := S \cup \{v\}$.

3. [*Termination test*] The computation terminates successfully if **P** contains all vertices of the graph, and if a circuit is being sought, then moreover the first and last vertex of **P** must be connected by an edge. If these conditions are not fulfilled, then either go to step 2, if during the last performance of step 2 the path **P** was being prolonged, or continue to step 4, if **P** was not prolonged.

4. [*Modify the path*] If the path **P** is formed by the sequence $v_0, ..., v_k$ and if there exists a number i with $1 \leq i \leq k - 2$ such that $\{v_i, v_k\}$ is an edge of the graph **G** and $v_{i+1} \notin S$, then change **P** to the following sequence

$$v_0, ..., v_{i-1}, v_i, v_k, v_{k-1}, ..., v_{i+2}, v_{i+1},$$

and go to step 2. If no such i exists, the computation terminates without having found a hamiltonian path or circuit. ●

The above algorithm proceeds by trying to prolong the given path **P** directly, and if this is impossible, it tries to modify **P** in such a way that it could be further prolonged from the new end-vertex. The manner in which

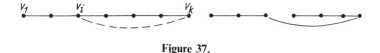

Figure 37.

the modification is performed is indicated in figure 37. By using the set S we prevent the modification from resulting in an endless cycle.

Practical experiments with Pósa's algorithm are described in [141]. Random edges were inserted into a set of 500 vertices until a connected graph without vertices of degree 1 was obtained (a necessary condition for the existence of a hamiltonian circuit). Then Pósa's algorithm was applied, and in 57 out of 60 experiments, a hamiltonian circuit was found on the first trial, and in the remaining cases it turned out to be sufficient to re-label the vertices and use the algorithm once again in order to obtain a hamiltonian circuit. Thus, the algorithm was successful in all the cases. This agrees with the theoretical results given by Pósa in his paper [180]. We will turn to them in §10.5.

9.3 The travelling salesman problem

We are going to present two approaches to the solution of the travelling salesman problem. The first one is based on the method of penalisation of vertices, and the experimental results found in the literature show that the method can be used to find the optimum solution of the travelling salesman problem for numbers of vertices up to 60 or 70. The latter one is the heuristic method of Lin and Kernighan [191] which does not always yield the optimum solution but makes it possible to process substantially larger graphs. Further methods, e.g. those described in [138] and [172], can be found in the survey paper [72]. Throughout the present section, the vertices of the graph under study are denoted by $v_1, ..., v_n$.

The method of penalisation is based on the following lemmas 9.3.1 and 9.3.3.

Lemma 9.3.1. *A subgraph H of a graph G is a hamiltonian circuit if, and only if, it has exactly n vertices and n edges (where n is the number of vertices of the graph G), it is connected, and each of its vertices v fulfils $\deg_H(v) = 2$.*

Proof. The necessity is obvious. Suppose, conversely, that H satisfies the last condition. Then H is a circuit or a system of circuits. However, the first condition excludes the possibility that H is a system of more than one circuit. ●

By leaving out the statement on the degrees in H from the above conditions, we obtain the following definition.

Definition 9.3.2. A subgraph of a connected graph G is called a 1-*tree* if it contains all vertices of G, it is connected, and its number of edges is equal to its number of vertices. If the graph G is edge-labelled, then a 1-tree with the minimum sum of the labellings of edges is called a *minimum* 1-tree. ●

We know that a tree of n vertices contains $n - 1$ edges. Therefore, a 1-tree is obtained by adding one edge to a tree; in other words, we admit the existence of a single circuit. A minimum tree, then, can be obtained by adding to a minimum spanning tree of the graph an edge not lying in the spanning tree with as small a labelling as possible. The algorithms of §6.2 present a simple and fast procedure for finding a minimum 1-tree. For example, when proceeding according to algorithm 6.2.2, we insert the first of the edges including a circuit into the constructed graph. A hamiltonian circuit is a special case of a 1-tree, and therefore, the sum of labellings of the edges of a minimum 1-tree is a lower bound on the value of the solution of the corresponding travelling salesman problem.

Lemma 9.3.3. *Denote by c_{ij} the labelling of the edge $\{v_i, v_j\}$ in the edge-labelled graph G, and let $t_1, ..., t_n$ be given numbers. By changing* the labellings in G from c_{ij} to $c_{ij} + t_i + t_j$ (for all vertices $\{v_i, v_j\}$ of G), the solution of the travelling salesman problem remains the same, although minimum 1-trees can be changed.

Proof. Each vertex the graph is incident on precisely two edges of any hamiltonian circuit. The above change of the labelling of edges thus leads to a change of the labellings of all hamiltonian circuits by the same number, namely, the double of the sum of the numbers t_i. Nothing like that holds for 1-trees. ●

The numbers t_i are called *penalisations of vertices*. Penalisation does not change the required solution of the travelling salesman problem, but we can try to change the form of a given minimum 1-tree in such a way that it becomes a circuit — then it will represent the solution we search, see 9.3.3.

As a measure of how much a given 1-tree T differs from a circuit, we are going to consider the number $D(T)$ equal to the sum of the absolute values of the expressions $\deg_T(v_i) - 2$ ranging over all vertices of the graph. In fact, according to lemma 9.3.1, a 1-tree T is a circuit if, and only if, $D(T) = 0$. If $\deg_T(v_i) > 2$ and we choose a positive penalisation of the vertex v_i, then the labelling of the edges incident with v_i will increase, and hence we can expect that as a consequence of penalisation those edges will be deleted from the minimal 1-tree, which will lead to a decrease of the number $\deg_T(v_i) - 2$. If $\deg_T(v_i) = 1$, we choose, conversely, a negative penalisation in order to increase the degree of the vertex. Although penalisation can influence the degrees of vertices in the 1-tree in a substantially different way than intended, practical experiences with the method have been satisfactory.

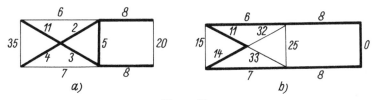

Figure 38.

An example of the influence of penalisation on the form of the minimal 1-tree is shown in figure 38 where the penalisation $t_i = 10(\deg_T(v_i) - 2)$ is used. The edges of the 1-tree before penalisation are denoted by bold segments in the left-hand figure, and those after penalisation are denoted by bold segments in the right-hand one. The result of penalisation in that case is excellent: a minimum hamiltonian circuit has been found.

Algorithm 9.3.4. A general scheme of the method of penalisation
Input data: a graph G with vertices labelled by numbers c_{ij}.
Task: find a minimum hamiltonian circuit in G.
Auxiliary variables: for each vertex v_i, a penalisation t_i and a number z_i
are used.

1. [*Initiation*] For $i = 1, ..., n$ put $t_i := 0$.
2. [*Determine a minimum 1-tree*] Determine a minimum 1-tree T in the
graph G with labellings changed from c_{ij} to $c_{ij} + t_i + t_j$ for all edges.
3. [*Termination test*] If T is a circuit, the computation terminates,
returning the result T.
4. [*Change penalisation*] For each $i := 1, ..., n$ choose a number z_i
and put $t_i := t_i + z_i$. Go to step 2.

It can happen that the above procedure repeats the cycle formed by
steps 2, 3 and 4 indefinitely without finding a solution, be it for fundamental
reasons or because of an unsuitable choice of the numbers z_i. Consequently,
it is necessary to add to step 3 a mechanism which terminates the algorithm
in such a case, or else we must rely on the operator or the computer system
to perform the termination after the assigned running time has been used up.

We have already encountered the simplest way of choosing the numbers z_i.

Algorithm 9.3.5.
Choose a number $c > 0$, and proceed according to algorithm 9.3.4
defining the numbers z_i in step 4 by $z_i := c(\deg_T(v_i) - 2)$. ●

The strategy of the last algorithm does not lead to a solution quite so
often as the above example may indicate. A better procedure based on
determination of z_i by the method of the largest descent is described in the
paper [148], and the reader can find further possibilities in the book [10],
where the question is studied in detail. Sometimes, however, even the most
perfect choice of the number z_i does not make it possible to find the circuit
being sought by the penalisation method. In that case, methods of § 4.3 are
applied as follows.

Given two disjoint sets X and Y of edges of the given graph, we denote
by $T(X, Y, t_1, ..., t_n)$ a 1-tree which contains all edges lying in the set X,
no edge from the set Y and, subject to those restrictions, has the smallest
sum of labellings of edges with respect to the penalised labelling
$c_{ij} + t_i + t_j$. For an unsuitable choice of X and Y, such a 1-tree does not
exist, of course.

By an admissible partial solution is meant a pair of sets X and Y. The
computation proceeds as follows.

We first try, by the procedure of 9.3.4, to modify the form of the 1-tree $T(\emptyset, \emptyset, t_i)$ by a change of penalisation in such a way that it becomes a circuit. This first phase of computation is nothing more than a performance of algorithm 9.3.4. The general phase of computation proceeds, provided that we have not managed to find in the assigned time a penalisation which would change $T(X, Y, t_i)$ into a circuit, by choosing an arbitrary edge h outside $X \cup Y$. We often repeat the whole computation with respect to both the pair $X \cup \{h\}$ and Y and the pair X and $Y \cup \{h\}$. The computation is thus split into two branches, and in each of them we try to find the solution by first using penalisation and then, eventually, subdividing to two branches again. The final solution for given X and Y is then equal to the better of the solutions for $X \cup \{h\}$, Y and X, $Y \cup \{h\}$.

We again see an upper bound on the total solution which is compared with the labelling of the 1-trees $T(X, Y, t)$; those yield a lower bound on the labelling of the hamiltonian circuit containing all edges of X and no edges on Y. The computation in a branch is, naturally, terminated also when for the chosen values of X and Y the tree $T(X, Y, t_i)$ does not exist, which happens if either X itself contains two circuits, or, conversely, the graph is no longer connected after deleting Y from it.

Details of implementation of a procedure of the above type can be found, for example, in [148] which also describes practical experiments using the choice of z_i presented in algorithm 9.3.5: randomly distributed points of a square region with the distance determined by the usual euclidean metric were processed in 0.16 minutes when $n = 22$ and in 4.3 minutes when $n = 64$. To mention another interesting experiment, a closed path of a chess knight through all the squares of the chess board was found in 6.93 minutes.

The procedure designed by S Lin and B W Kernighan [191] also searches for a solution of the travelling salesman problem. It attempts to find the circuit being sought in a different way: by a re-connection of edges, the aim of which is to decrease the total sum of the labellings of edges. To simplify the formulation, we choose as the basic object not a circuit, but a sequence of vertices $w_1, \ldots, w_n, w_{n+1}$ such that $\{w_i, w_{i+1}\}$ are edges, the sequence $w_1, \ldots, w_n$ is one-to-one, and w_{n+1} is equal to one of the vertices $w_1, \ldots, w_n$. Figure 39 shows that this object is either a 'circuit with a tail', or, if

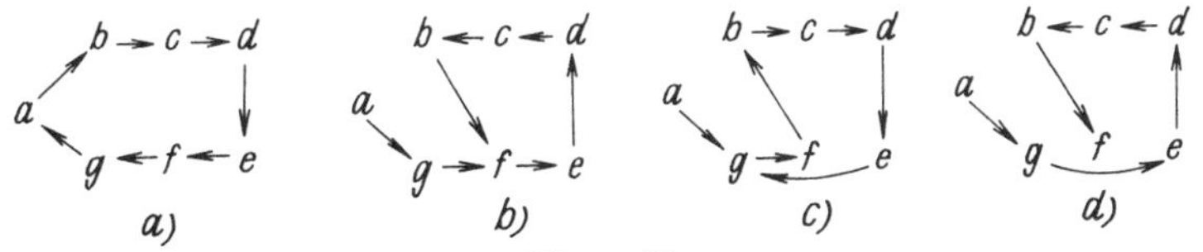

Figure 39.

$w_{n+1} = w_1$, it is just a circuit as a special case. The edges of the object are considered to be directed.

The modification will be performed similarly as in Pósa's algorithm above. From the sequence $w_1, ..., w_{n+1}$, in which the edge $\{w_n, w_{n+1}\}$ is viewed as an incomplete closure of the path $w_1, ..., w_n$ into a circuit, we construct a sequence as follows:

$$w_1, ..., w_i = w_{n+1}, w_n, w_{n-1}, ..., w_{i+1}$$

and then we choose a vertex w_j such that $j \neq i, i+1, i+2$ to supplement the above sequence to the following one:

$$w_1, ..., w_i, w_n, w_{n-1}, ..., w_{i+1}, w_j.$$

We assume here, of course, that $\{w_{i+1}, w_j\}$ is an edge of the given graph. This method is usually applied (and will be so formulated) for the study of a labelled complete graph, where for an edge connecting two vertices to exist it is sufficient that the vertices be distinct. An example of several modifications of that type is shown in figure 39.

The difference between the labelling of the edge $\{w_i, w_{i+1}\}$ (which is to be deleted) and that of $\{w_{i+1}, w_j\}$ (to be added) is called the *gain*. The principle of the Lin–Kernighan procedure also admits, in order to overcome local minima more easily, modifications with a negative gain; however, the changes will be performed in a series in which the total sum of gains obtained is positive.

We now present a description of the method. The labelling of the edge with end-vertices x and y will be denoted by $c(x, y)$.

Algorithm 9.3.6. Travelling salesman problem
Input data: edge-labelled complete graph K_X the set of vertices of which is denoted by X.
Task: find in K_X a hamiltonian circuit of as small a sum of labellings of edges as possible.
Auxiliary variables:
  P, a sequence of $n + 1$ vertices of the graph,
  T, the largest improvement in the course of the series of modifications,
  C, a circuit realising the improvement,
  Z, the sum of gains in the course of the series of modifications,
  X and Y, sets of edges to be deleted and added, respectively, in the given series.
    1. [*Choose the basic circuit*] In the graph K_X choose an arbitrary hamiltonian circuit C.
    2. [*Choose the basic sequence*] Construct a sequence $P = (w_1, ..., w_n, w_{n+1})$

from the circuit C in such a way that $w_1 = w_{n+1}$ and for each $i = 1, \ldots, n$, the edge $\{w_i, w_{i+1}\}$ lies on the circuit C. (We can choose w_1 arbitrarily, then w_2 is one of the two neighbours of w_1 on the circuit, and this determines the other members of P uniquely.)

3. [*Initiate the series of improvements*] Put $T := 0$ and $Z := 0$.

4. [*First modification*] Choose j with $1 < j < n - 1$ and $c(w_1, w_n) > c(w_j, w_n)$ and put

$$P = (w_1, \ldots, w_n, w_j);$$
$$Z := c(w_1, w_n) - c(w_j, w_n);$$
$$X := \{w_1, w_n\}; \quad Y := \{w_j, w_n\}.$$

If no such j exists, go to step 9.

5. [*Search further modification*] Let $P = (z_1, \ldots, z_n, z_i)$. Choose k subject to the following conditions:

(i) $\{z_{i+1}, z_k\} \in X$ and $\{z_i, z_{i+1}\} \notin Y$ (we are not going to add a vertex which has previously been deleted, nor delete one which has previously been added),

(ii) $Z + c(z_i, z_{i+1}) - c(z_{i+1}, z_k) > 0$ (the total gain of the series must be positive),

(iii) $\{z_k, z_{k+1}\} \in Y$ (in order not to get into conflict with (i) in the next modification).

If no such k exists, go to step 9.

6. [*Actualise Z*] Put $W := Z + c(z_n, z_1) - c(z_n, z_i)$ (which is the difference of labelling of the initial circuit of the present series and the circuit we would obtain if we were to close P immediately by re-connecting the edge leading from the vertex z_n). If $W > T$, put $T := W$ and $C :=$ the circuit of edges $\{z_1, z_2\}, \ldots, \{z_{n-1}, z_n\}, \{z_n, z_1\}$.

7. [*Modify*] Put

$$P := (z_1, \ldots, z_i, z_n, z_{n-1}, \ldots, z_{i+1}, z_i)$$

and add the edge $\{z_i, z_{i+1}\}$ to X and the edge $\{z_{i+1}, z_k\}$ to Y.

8. [*Test of continuation of the series of modifications*] If $Z > T$, go to step 5.

9. [*Test of continuation of computation*] If $Z > 0$, go to step 3 (and try improving the best circuit yet found, which is C, in a further series of modifications); else, terminate the computation. ●

The original algorithm described in [191] is more extensive than that of 9.3.8 since in the case where step 9 shows that no improvement has been achieved, each of the choices of w_1 and w_2 in step 2 is processed separately, and for each of them all the possibilities of improvement by a choice of j in step 4 and the final choice of k in step 5 are considered. When even that does not lead to an improvement, the authors give further possible modifications which are admissibly performed on the sequence **P**. All these adjustments increase the probability of finding the optimal solution, but they prolong the computation, and it thus depends on the relationship of the requirements on speed and simplicity on the one side and the quality of solution on the other side whether or not we decide to make use of them. Experimental results obtained on the GE635 computer have also been given. With all the above adjustments, the algorithm was capable of processing a problem of n vertices randomly distributed in a square region with the usual euclidean distance in approximately 8 seconds for $n = 50$ and in 25 seconds for $n = 110$. In contrast to the exact algorithms described above, it is not certain that the procedure under consideration will find an optimum solution. By a repeated performance of the procedure with a random choice of the basic circuit C it was established that the best of the results obtained in all experiments (which need by no means be the real optimum solution) was achieved in approximately 89 per cent of cases if $n = 50$, and 21 per cent of cases for $n = 110$. It is thus advisable to repeat the computation several times, which will correspondingly increase the required running time, of course.

9.4 Isomorphism of graphs

There exist classes of graphs within which it is not too time-consuming to decide whether two graphs are isomorphic or not (although the corresponding algorithms are often very complicated). As an example, let us mention planar graphs which we have investigated in §6.8, further, graphs with a bounded degree of vertices [190] or with bounded multiplicity of the eigenvalues of the incidence matrix [51]. The general problem of isomorphism of graphs is much more difficult; moreover, its NP-completeness (though some isomorphism related problems are NP-complete [195]) has not yet been proved in spite of a great effort in this direction. It is only known that the problem is equivalent to the problem of searching for isomorphism of other mathematical objects, e.g. some special types of graphs, semigroups, automata, etc; see for example [64].

The first step to perform when we are faced with the problem of searching for isomorphism of two graphs is to verify the equality of the basic numerical characteristics of the graphs, in the first place the number of vertices and edges. Further invariants can be deduced from the incidence matrix which is not an invariant itself, since it depends on the ordering of vertices with respect to which it has been determined, but which makes it possible to determine, for example, the characteristic polynomial of the matrix, the roots of the polynomial, the eigenvectors normalised to unit length, the determinant, and further numbers which must be equal for two isomorphic graphs. Those numbers cannot, unfortunately, be used for the proof of the existence of an isomorphism since pairs of non-isomorphic graphs are known for which all the mentioned characteristics are equal.

A survey of methods for searching for isomorphism of graphs is presented in the paper [223]. The best-known method is based on a division of vertices into classes according to degree. Thus, when an isomorphism takes a vertex x of one graph onto the vertex y of the other graph, then the degrees of x and y in the corresponding graphs must be equal. A further refinement can be achieved by means of degrees of higher order. Denote, for example, for a given p by $\deg_{G,p}(v)$ (or for short, $\deg_p(v)$), the number of vertices of degree p which are connected by an edge with the vertex v. An isomorphism can assign to a given vertex v only such a vertex which has the same value of the corresponding vector $(\deg_0(v), \ldots, \deg_n(v))$.

As an example, we present a search for isomorphism of the graphs of

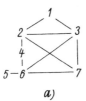

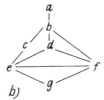

a) Figure 40. b)

figure 40. They have seven vertices each, and hence, the brute-force approach would have to go through $7! = 5040$ possible bijections, which is possible, but unnecessary. We can divide the vertices according to degree as follows:

degree 4: 2, 3, 6 b, e, f,
degree 3: 7 d,
degree 2: 1, 4 c, g,
degree 1: 5 a.

This alone forces the correspondences 3–d and 5–a. The number of candidates for an isomorphism has thus been reduced to $3! \cdot 1! \cdot 2! \cdot 1! = 12$. The search of all these possibilities would be easy, but we are going to continue the subdivision of vertices. We divide the vertices according to their vectors $(\deg_1(v), \ldots, \deg_4(v))$, no longer considering the vertices 5, 7, d, a.

$(0, 1, 1, 2)$: 3 f,
$(0, 2, 1, 1)$: 2 e,
$(1, 1, 1, 1)$: 6 b,
$(0, 0, 0, 2)$: 1, 4 c, g.

This yields further correspondences 3–*f*, 2–*e*, and 6–*b*, and it only remains to consider the remaining two candidates for isomorphism. One of them can be excluded by the fact that the vertices 1 and *g* are connected with a vertex of type $(0, 1, 1, 2)$, whereas the vertices 4 and *e* are not. Thus, we get the correspondences 1–*g* and 4–*c*. It remains to check that we have actually obtained an isomorphism, and the problem is solved. Neither the above procedure, nor the improvement described in [76], have to yield a complete division of the vertices and thus form a unique correspondence of the vertices of the graphs. We thus may have to supplement the procedure by considering all the remaining possibilities. This is the idea of algorithm 9.4.2 below, for the formulation of which we will use the following concept.

Definition 9.4.1. A quasi-ordering of a set X is a binary relation $\prec$ such that there exists a partition of the set X into pairwise disjoint subsets $X_1, ..., X_k$ with the property that given arbitrary members $x \in X_i$ and $y \in Y_j$, then we have $x \prec y$ if, and only if, $i < j$. ●

Algorithm 9.4.2. Isomorphism of graphs
Input data: graphs G and H.
Task: find an isomorphism of the graphs G and H, or prove that it does not exist.
Auxiliary variables: quasi-orderings $\prec_G$ and $\prec_H$ of the set of vertices of G and H, respectively, determining disjoint decompositions $X_1, ..., X_k$ and $Y_1, ... Y_k$, respectively.

1. [*Initiation*] At the beginning, the relations $\prec_G$ and $\prec_H$ have unique classes X_1 and Y_1, respectively; in other words, neither $p \prec_G q$, nor $p \prec_H q$ hold for any pair p and q of vertices.

2. [*Refine the partitions*] Perform steps 2a to 2c while the quasi-orderings $\prec_G$ and $\prec_H$ are actually being refined in the course of those steps:

2a. [*Determine degrees*] For each vertex x of the graph G determine the numbers $\deg_G(x, X_i)$ for $i = 1, ..., k$. Analogously, determine the numbers $\deg_H(y, Y_i)$, $i = 1, ..., k$ for each vertex y of H.

2b. [*Refine the quasi-orderings*] Refine the relation $\prec_G$ by putting $x \prec_G x'$ if, and only if, either $x \prec_G x'$ has been true before, or x and x' were previously lying in the same class of the relation $\prec_G$, and there exists an i such that

$$\deg_G(x, X_j) = \deg_G(y, X_j) \quad \text{for } j = 1, ..., i - 1$$

and

$$\deg_G(x, X_i) < \deg_G(y, X_i).$$

The relation $\prec_H$ is refined analogously. This changes, in general, both the form and the number of the sets X_i and Y_i.

2c. [*Test the coincidence*] If the newly defined relations $\prec_G$ and $\prec_H$ have a different number of classes, or if two corresponding classes have a different number of elements, the computation terminates since G and H are not isomorphic.

3. [*Consider the remaining possibilities*] Construct successively all bijections which take vertices of the graph G onto those of the graph H in such a way that the class X_i of the relation $\prec_G$ is mapped onto the corresponding class Y_i of the relation $\prec_H$. If none of those bijections is an isomorphism, then the graphs G and H are not isomorphic. ●

If we apply algorithm 9.4.2 to regular graphs, i.e. graphs in which all degrees of vertices are the same, then step 2 is incapable of distinguishing any pair of vertices, and hence, in step 3 we would have to consider all the $n!$ possible bijections, which is impossible even in case of comparatively small graphs. In that case we can proceed by choosing a vertex x of the graph G and for each vertex y of the graph H we try and find an isomorphism taking x to y. This causes a non-symmetry in the isomorphism problem: if we choose the initial quasi-orderings in such a way that their classes are $\{x\}$ and $\{y\}$, respectively, and the set of all the remaining vertices of the graphs G and H, respectively, then in step 2 the degrees determined in 2a are different and hence, with a bit of luck, we will manage to get a sufficiently fine partition at the end of the computation. Without luck, we will have to continue in the same manner. Combining the methods of §4.3 with the above algorithm we obtain an algorithm which is, in a majority of cases, capable of finding the isomorphism of graphs of rather large sizes sufficiently quickly.

10

Probabilistic analysis of algorithms

In contrast to all the preceding chapters, in this one we are going to show that a number of procedures we have studied so far are not as bad it would seem from their worst-case behaviour. We will analyse the algorithm Quicksort, the search in a binary search tree, the primality test, and some algorithms processing graphs.

10.1 Introduction

In the previous chapters we have evaluated algorithms according to their behaviour in the worst case. When we are able to prove that a certain algorithm is satisfactory even in the worst case, the result is valuable also from the practical point of view. However, when we show that, conversely, in some situations the computation would take so much time that it could not be practically performed, or would not give a sufficiently precise result, this does not mean that the algorithm is worthless. For it is possible that an unfavourable type of input data is quite rare, and that in the common cases the computation proceeds quickly and with a satisfactory result.

There are large numbers of examples of algorithms which are considered to be very good in spite of their non-satisfactory behaviour in the worst case. The best known of them is the simplex method for the problem of linear programming. Empirical experience tells us that the length of computation of this method is proportionate to the size of the processed problem, while artificial examples show that the correspondence can even be exponential; see [177] and [273]. Analogously, the algorithm Quicksort is considered to be one of the quickest sorting algorithms in spite of the fact that in an unsuitable case n items are processed in running time proportionate to n^2, whereas there are several algorithms working in running time $O(n \log n)$.

It is thus quite possible that the NP-complete problems we have studied in Chapter 7 also have a solution that is satisfactory from the probabilistic point of view. This impression is supported by the fact that the proofs of NP-completeness (which are considered to be proofs that the corresponding problems are difficult) are often so complicated, that the graphs constructed in them cannot be expected to appear in practical problems. In subsequent sections of the present chapter we are going to show that this impression is quite justified.

We will now show by several examples how a probabilistic evaluation of algorithms is performed. Let us first consider the algorithm Quicksort, the aim of which is to sort given numbers $x_1, ..., x_n$. It is clear from the above description of the algorithm that the performance of computation does not depend on the size of the numbers to be processed, but solely on their interrelationship; for example, the following sequences: $0, 5, 1, 2$ and $10, 13, 11, 12$, will be processed in the same way since in both of them we have $x_1 < x_3 < x_4 < x_2$. The number of different performances of computation is thus equal to the number of permutations on the input numbers. If we determine for each of those $N!$ permutations the length of computation, and then compute the arithmetical mean, we will obtain a number which can be considered to be the average running time necessary to sort n numbers. In §10.2 we will show that this number is proportionate to $n \log n$, which shows that the average speed of Quicksort is substantially higher than that of the worst case.

Suppose now that we want to determine the average behaviour of some approximate algorithm A for colouring of graphs, e.g. some of the algorithms of §9.1. We usually proceed by determining, for each n and each of the $2^{\binom{n}{2}}$ graphs G on an n-point set, the ratio $A(G)/\chi(G)$ where $A(G)$ is the number of colours assigned to G by the algorithm A, and $\chi(G)$ is the optimum number of colours. Then we compute the arithmetical mean over all values corresponding to a given n-point set. If the resulting number is close to 1, we can say that the algorithm is suitable on average.

The above example shows the weak point of the probabilistic evaluation by average behaviour: if we are, for example, only interested in the class of graphs not containing a circuit of length 3, then it may happen that unfavourable behaviour of the investigated algorithm working in that class is covered by its good behaviour for graphs which form a majority, but which are of no interest to us. Then it is necessary to perform the full analysis all over again, and to determine the average only over the graphs which we are going to encounter. Thus, in contrast to the absolute validity of the worst-case analysis of behaviour, the analysis of the average behaviour is always restricted to a certain class of input data, or more generally, to a certain probability distribution on them.

There is yet another approach to the solution of difficult problems which gives a high probability of success. As an example we consider the decision as to whether a given number is a prime. We will give in §10.6 an assertion

$\mathbf{T}(n, k)$ which has the property that a number n is a prime if, and only if, $\mathbf{T}(n, k)$ does not hold for any $k = 1, ..., n$. Moreover, it has been proved that whenever n is not a prime, then the number of ks such that the assertion $\mathbf{T}(n, k)$ holds is at least $n/2$.

The above facts can be used as follows: given n, choose randomly m numbers $k_1, ..., k_m$, and verify the assertion $\mathbf{T}(n, k_i)$ for $i = 1, 2, ..., m$. If one of the assertions holds, then we know for sure that n is not a prime. If none of them holds, we may state that n is a prime, and the probability that we are making an error is 2^{-m}, which is very small.

In general, we may try to construct an algorithm which in some steps can decide for one of several possible continuations in a random way (e.g. by using a generator of random numbers). We admit that in some cases the result will be unsatisfactory, or even wrong. When evaluating the average behaviour we do not form the average over all possible input data of a given size, but over all possible computations determined by the unique given input data. The result of such analysis is then independent of the frequency with which different input data are encountered.

The above algorithm for determining the primality of numbers is designed in such a way that it answers either 'not' (the given number is not a prime) and the result is certainly true, or 'yes', and then there is a non-zero probability, however small, that the answer is wrong. More suitable are thus the algorithms which reply 'yes', 'no' or 'don't know', and the first two answers are certainly true whereas the probability of the third one is extremely small.

This approach is very promising, and some theoretical results have already been presented; see [43, 44 and 132]. On the other hand, at the present time only a few algorithms are known which solve problems in this way that are not solvable by classical methods. For more information the reader is referred to [268].

10.2 Sorting

In the present section we will concentrate on the analysis of the average speed of computation of comparison algorithms for sorting, and we will prove that for the average time required to process n items we have the estimate $\Omega(n \log n)$ — the same as for the worst case. We are going to assume that the task of the algorithm is to sort a given permutation of numbers, and that the probability of processing is the same for all permutations. For the algorithm Quicksort we will then show that, in contrast to the time estimate $O(n^2)$ for the worst case, the estimate in the average case is $O(n \log n)$, which agrees with the empirical experience according to which the algorithm belongs (as its name indicates) to the quickest known algorithms for sorting.

Theorem 10.2.1. *Let A be a comparison algorithm for sorting. For each permutation p on the set of numbers $1, \ldots, n$ we denote by $t(p)$ the number of steps of computation performed by the algorithm A processing the permutation p. Then we have*

$$\frac{1}{n!} \sum t(p) \geq \log n! = \Omega(n \log n),$$

where the sum is considered over all permutations p on the set $1, \ldots, n$.

Proof. By a distinguishing system we will mean a set M of sequences of 0s and 1s such that for two arbitrary distinct members $a_1, \ldots, a_n$ and $b_1, \ldots, b_m$ of M there exists $i \leq \min(m, n)$ with $a_i \neq b_i$. We further denote by $\|M\|$ the sum of lengths of all the members of M, and for each natural number m by $f(m)$ the minimum value of $\|M\|$ for all distinguishing systems of m sequences. We will prove by induction that

$$f(m) \geq m \log m.$$

It is obvious that $f(1) = 0 = 1 \log 1$. Suppose that $m > 1$ and the assertion holds for all smaller values. Choose a distinguishing system M of m sequences such that $\|M\| = f(m)$. Denote by M_0 and M_1 the set of all sequences in M with the first element 0 and 1, respectively. Let N_0 be the set of all sequences obtained from the sequences in M_0 by deleting the first 0, and analogously let N_1 be the set obtained from M_1 by deleting the first 1. It is easy to verify that both N_0 and N_1 are distinguishing systems. If M_0 has i elements, then N_0 also has i elements, and both M_1 and N_1 have $m - i$ elements. Thus, by induction, we get

$$\begin{aligned}
f(m) = \|M\| &= (i + \|N_0\|) + (m - i + \|N_1\|) \\
&\geq m + f(i) + f(m - i) \\
&\geq m + i \log i + (m - i) \log(m - i).
\end{aligned}$$

It is easy to prove that the function $i \log i + (m - i) \log(m - i)$ attains its minimum at $i = m/2$, and thus

$$\begin{aligned}
f(m) &\geq m + 2(m/2) \log(m/2) = m + m(\log m - 1) \\
&= m \log m.
\end{aligned}$$

Now, let A be a comparison algorithm for sorting, understood as a program of a random access machine. For each permutation p denote by $P(p)$ the following sequence of 0s and 1s: length of $P(p)$ is equal to the number of steps of computation necessary to process the permutation p and the ith coordinate of $P(p)$ is equal to 1 if, and only if, in the ith step of computation a conditional jump instruction was executed and the jump was actually performed. For two distinct permutations p_1 and p_2 the computations which process them must also be distinct, and hence, according to definition 5.3.2 we must have $P(p_1) \neq P(p_2)$. Therefore, the system M of all sequences $P(p)$ where p is a permutation on the numbers $1, ..., n$ is distinguishing, and since it has $n!$ members and since the length of the sequence $P(p)$ is equal to the number $t(p)$, the above inequality yields the following one

$$\sum t(p) = \|M\| \geq n! \log(n!) . \quad \bullet$$

In contrast to most of the important algorithms for sorting which work in running time $O(n \log n)$, the algorithm Quicksort requires, for some cases, running time proportional to n^2. We will now show that in the average case, however, the time $O(n \log n)$ is also sufficient for Quicksort.

Theorem 10.2.2. *Consider algorithm 5.2.1 in which the division point k is chosen as a key of the item lying in the middle of the processed sequence. For each permutation p of the members $1, ..., n$ denote by $t(p)$ the running time required by that algorithm for sorting p. Then for a suitable constant c we have*

$$\frac{1}{n!} \sum t(p) \leq cn \log n,$$

where the sum is considered over all permutations on the set $1, ..., n$.

Proof. The algorithm under consideration works, roughly speaking, as follows: it chooses a certain number i, divides the numbers $1, ..., n$ into three groups Z_1, Z_2 and Z_3 according to whether the numbers are larger than, equal to, or smaller than i, respectively, and then, by recursive application of itself, it sorts the sets Z_1 and Z_3, giving as a result the sorted group Z_1 followed by the group Z_2 and the sorted group Z_3. The group Z_2 consists, in our case, of the number i alone. It is easy to see that each permutation on the numbers $1, ..., i - 1$ appears with the same probability as Z_3, and each permutation of $i + 1, ..., n$ appears with the same probability as Z_1. Moreover, each of the numbers $0, 1, ..., n - 1$ has the same probability of being equal to the number of elements of Z_1. Denoting by $T(n)$ the mean time required to process the numbers $1, ..., n$, then we have $T(0) = T(1) = 0$, and for $n \geq 2$ the number $T(n)$ is the arithmetical mean of the numbers

$$c_1 n + T(i) + T(n - 1 - i)$$

for all $i = 0, 1, ..., n - 1$, where c_1 is a constant chosen in such a way that $c_1 n$ expresses the time required to partition the numbers into the groups Z_1, Z_2 and Z_3, to unify them and to perform other auxiliary operations. The numbers $T(i)$ and $T(n - 1 - i)$ describe the time required for the recursive applications of our algorithm to Z_1 and Z_3, respectively. Thus, we have

$$T(n) \leq c_1 n + \frac{1}{n} \sum_{i=0}^{n-1} [T(i) + T(n - 1 - i)] = c_1 n + \frac{2}{n} \sum_{i=0}^{n-1} T(i).$$

We will prove that for all $n \geq 2$ we have $T(n) \leq cn \ln n$, where $c = 2c_1$ and $\ln n$ denotes the natural logarithm (of base e). For $n = 2$ the assertion follows from the above inequality, and if $n > 2$, then

$$\sum_{i=2}^{n-1} i \ln i \leq \int_2^n x \ln x \, dx$$

$$= \left[\frac{x^2}{2} \ln x - \frac{x^2}{4} \right]_2^n \leq \frac{n^2}{2} \ln n - \frac{n^2}{4},$$

and hence,

$$T(n) \leq c_1 n + \frac{2c}{n} \left(\frac{n^2}{2} \ln n - \frac{n^2}{4} \right)$$

$$= \left(c_1 - \frac{c}{2} \right) n + cn \ln n = cn \ln n. \quad \bullet$$

10.3 Searching in dictionaries

The aim of this section is to give some results concerning the probabilistic analysis of algorithms used to perform operations on the binary search tree. An advantage of these algorithms is their simplicity, but in contrast to the more complicated methods, the running time required to perform the operations is in the worst case proportional to the number of elements of the tree. It is surprising that, in spite of the simplicity of these algorithms, their complete probabilistic analysis has not yet been performed; see [129]. On the other hand, a number of partial results are known, and we shall give some of these.

We will first estimate the running time necessary to store successively n items into an originally empty tree by the operation *INSERT* according to algorithm 5.3.3. In an unfavourable case, e.g. if the items are stored in an ordered sequence, we need the following total number of comparisons of keys:

$$0 + 1 + \ldots + (n - 1) = \frac{1}{2}n(n - 1).$$

However, for the average running time we have the following result.

Theorem 10.3.1. *Let* $t(p)$ *denote the running time necessary to store* n *numbers* $1, \ldots, n$ *into a binary search tree by the operation INSERT according to algorithm 5.3.5, assuming that they are stored in the ordering given by the permutation* p. *For a suitable constant* c *we then have*

$$\frac{1}{n!} \sum t(p) \leqq cn \log n,$$

where the sum ranges over all permutations p *on the numbers* $1, \ldots, n$.

Proof. The running time necessary to perform the operation *INSERT* is proportionate to the number of comparisons of keys of items which in the present case are the numbers to be stored. Denote by $T(n)$ the average number of comparisons which are performed in the course of storing randomly ordered numbers $1, \ldots, n$. It is clear that $T(1) = 0$. Suppose $n > 1$. The first number to be stored becomes the root of the tree under construction. Suppose that there exist i numbers smaller than the root. In the course of storing further numbers we see that it is necessary to compare the numbers with the root and to store i numbers in a random sequence into the left subtree, and $n - 1 - i$ numbers into the right subtree. Thus, in this case we perform, on average, $(n - 1) + T(i) + T(n - 1 - i)$ comparisons. Since i takes all the values between 0 and $n - 1$, inclusive, with equal probability, we have

$$T(n) = \frac{1}{n} \sum_{i=0}^{n-1} [n - 1 + T(i) + T(n - 1 - i)],$$

from which the required estimate is obtained in an analogous way to that of the proof of 10.2.2. ●

An exact analysis of the formula shown in the proof of the preceding theorem would yield the following estimate of the number of comparisons of k in the average case: $T(n) \leqq kn \log n$, where k is the natural logarithm of 4, i.e. $k \doteq 1.386$.

We next analyse a similar problem. We have stored into an originally empty binary search tree n randomly ordered items by the operation

INSERT and we ask about the average running time of performance of the following operations:

— successful *MEMBER*: we search, according to 3.5.2, for an item which lies in the tree;

— unsuccessful *MEMBER*: we search, according to 3.5.2, for an item which does not lie in the tree;

— *INSERT*: we insert, according to 3.5.3, an item into the dictionary which does not lie there;

— *DELETE*: we delete, according to 3.5.5, an item from the dictionary, which lies there;

— *MIN* and *MAX*: we search for a minimum and a maximum according to 3.5.4.

It follows from the description of the corresponding algorithms that *INSERT* does not differ in practice from the unsuccessful *MEMBER* as far as speed is concerned. The operation *DELETE* is, even in the unfavourable case in which the item to be deleted has two sons, roughly as quick as the unsuccessful *MEMBER* searching an item of key value slightly smaller than that of the item to be deleted. Finally, it is not difficult to see that *MIN* and *MAX* require, on the average, running time equal to that of the successful *MEMBER*. We will thus concentrate on the analysis of the operation *MEMBER* alone.

Given a binary search tree of n vertices, then there are n possible performances of the operation *MEMBER* in the successful case corresponding to the vertices of the tree, and $n + 1$ performances in the unsuccessful case, corresponding to the $n + 1$ possible relations of the key of the inserted item to the n distinct keys of the items which already lie in the tree. When speaking about the average numbers T_{succ} and T_{unsucc} of comparisons necessary to perform the two variants of *MEMBER*, respectively, we mean the average of the number of comparisons performed by algorithm 3.5.2 in the corresponding n or $n + 1$ performances, respectively.

Lemma 10.3.2. $T_{unsucc} = \dfrac{n}{n + 1}(T_{succ} + 1).$

Proof. For each vertex v in the binary search tree, let h_v denote the number of comparisons of keys in the course of searching v by the operation *MEMBER*. This number corresponds to the depth of the vertex v. Consequently, we have

$$T_{succ} = \frac{1}{n}\sum h_v,$$

where the sum ranges over all vertices of the tree. Denote, further, by N_v

the number of sons of the vertex v, then $2 - N_v$ performances of the unsuccessful *MEMBER*, out of the total number $n + 1$, terminate at v. Consequently, we have

$$T_{\text{unsucc}} = \frac{1}{n + 1} \sum (2 - N_v) h_v,$$

where the sum ranges over all vertices again. We will prove by induction the following equality:

$$\sum (2 - N_r) h_v = n + \sum h_v.$$

In the case of a unique vertex, both sides are equal to 2. Suppose now that the assertion is valid for a certain tree S, and let us insert a new successor, denoted by z, of a vertex v of S. The right-hand side of the above equation will be increased by 1 in the first summand and by $h_z = h_{v+1}$ in the latter one, and hence the total increase will be $h_v + 2$. The number of sons will increase by 1 and hence $(2 - N_r) h_v$ will be decreased by h_v, but at the same time a new summand $(2 - N_z) h_z = 2(h_v + 1)$ will appear in the left-hand sum, and hence the sum will also be increased by $h_r + 2$. Consequently, the equality will remain valid, which concludes the proof by induction.

We thus have, as required, the following equality:

$$T_{\text{unsucc}} = \frac{1}{n + 1} \sum (2 - N_v) h_v = \frac{1}{n + 1} \left(n + \sum h_v \right)$$

$$= \frac{n}{n + 1} \left(1 + T_{\text{succ}} \right). \quad \bullet$$

The above equality, which states that the average running time of performance is almost the same for the successful *MEMBER* and the unsuccessful one, holds for an arbitrary shape of the binary tree. We are now ready to estimate the average running time necessary for a performance of *MEMBER* in a random tree.

Theorem 10.3.3. *Suppose that we insert n items in a random sequence into an originally empty binary search tree by the operation INSERT according to 3.5.3. If each permutation of the ordering of items is equally probable, then the average running time necessary for the subsequent performance of the operation MEMBER according to 3.5.2 is $O(\log n)$.*

Proof. In order to search for an item lying in the dictionary it is necessary to perform one more comparison of keys than for performance of the operation *INSERT* by which the item was stored into the tree. Thus, we have $T_{succ} = O(\log n)$, according to 10.3.1, and consequently, the same estimate is valid for T_{unsucc}, according to 10.3.2. ●

From the remark following theorem 10.3.1 and from the last proof above we can conclude that in a random dictionary the average number of comparisons of keys according to algorithm 3.5.3 is approximately 1.39 log n. In the optimum binary search tree in which all leaves lie in the last layer we have, for $h = T_{unsucc}$, the number of vertices equal to $2^h - 1$, and therefore, conversely, for a performance of *MEMBER* we need approximately log n comparisons of keys. The mean running time for performance of dictionary operations in a binary search tree is, assuming sufficient randomness, worse by roughly only 39 per cent than the optimum running time. This, however, has only been proved in cases where we measure the duration of the operation which follows the n initial insertions. Nevertheless, if the dictionary operations *INSERT* and *DELETE* alternate in a random way, then it has been shown by practical experiments that the average running times remain roughly the same. A theoretical analysis of that case is too complicated, which is why it has never been completely performed.

10.4 Random graphs

In § 10.1 we have already indicated the way in which one can determine the average behaviour of an algorithm processing graphs. Given an algorithm A, we first assign to each graph G a number $f(G)$ which describes in a certain sense the behaviour of the algorithm A applied to the graph G. If we are interested, for example, in the speed of computation, then we put $f(G)$ equal to the number of steps performed by the algorithm A when processing the graph G. When evaluating an approximate algorithm, we can denote by $f(G)$ the ratio of the values of the solution found by A to the optimum solution, and so on. Then for each natural number n we compute the arithmetical mean of the numbers $f(G)$ over all graphs G with a prescribed n-point set of vertices (usually the set $1, ..., n$), and we obtain a number describing the average behaviour of the algorithm A applied to n-vertex graphs. Since the exact determination of such numbers is usually difficult, we are often satisfied with an estimation of the asymptotic behaviour of their average understood as a function of n.

We have already mentioned that in cases where some graphs are encountered more often, some only rarely, and some not at all, then it can happen that the average of the numbers $f(G)$ does not describe the behaviour of the algorithm adequately. In that case we can improve the situation by introducing a function w called the *weight* of the graph G, which corresponds to the frequency with which we encounter the graph G. The numbers $w(G)$ are often chosen in such a way that their sum ranging over all graphs G or a given vertex set is equal to one; then $w(G)$ is directly the probability of occurrence of the graph G. The following number:

$$\frac{\sum w(G) f(G)}{\sum w(G)}$$

represents the quantitative expression of the average behaviour of the algorithm applied to n-vertex graphs; the sums range over all graphs with the vertex set $\{1, ..., n\}$. This number is called the *weighted mean* of the numbers $f(G)$.

With a few exceptions, we are only going to investigate, besides the (non-weighted) arithmetical mean, the following case: a function $m = m(n)$ is given, and for each graph G of n vertices we put $w(G) = 1$ when the number of edges of G is equal to m, else $w(G) = 0$. This will enable us to study the average behaviour of algorithms in dependence not only on n, but also on the density of the graph, i.e. the ratio of the numbers of edges and vertices.

In order to simplify notation and formulations of subsequent theorems, we introduce the following definition.

Definition 10.4.1. Let f be a function assigning to each graph G a real number $f(G)$, and let m be a self-map of the set of all natural numbers. Put

$$u(G) = 1 \quad \text{for each graph } G,$$

and

$$v(G) = \begin{cases} 1 & \text{if the number of edges of } G \text{ is } m(n) \text{ where } n \\ & \text{is the number of vertices of } G, \\ 0 & \text{else.} \end{cases}$$

Denote further

$$f(\mathbf{G}_n) = \frac{\sum u(G) f(G)}{\sum u(G)} \quad \text{and} \quad f(\mathbf{G}_{nm}) = \frac{\sum v(G) f(G)}{\sum v(G)}$$

where all the sums range over the set of all graphs with the vertex set $\{1, ..., n\}$. ●

Since the number of pairs in the set $\{1, ..., n\}$ is $\binom{n}{2}$, it is possible to construct $2^{\binom{n}{2}}$ graphs in that set of vertices, and $\binom{\binom{n}{2}}{m}$ of those will have m edges. This proves the following lemma.

Lemma 10.4.2.

$$f(\mathbf{G}_n) = \frac{\sum f(G)}{\exp\binom{n}{2}},$$

where the sum ranges over all graphs G with the vertex set $\{1, ..., n\}$, and

$$f(\mathbf{G}_{nm}) = \frac{\sum f(G)}{\binom{\binom{n}{2}}{m}},$$

where the sum ranges over all graphs G with the vertex set $\{1, ..., n\}$ which have $m(n)$ edges. ●

Definition 10.4.3. Let V be a certain property, and let $m = m(n)$ be a self-map of the set of all natural numbers. Put

$$f(G) = \begin{cases} 1 & \text{if the graph } G \text{ has the property } V \\ 0 & \text{else.} \end{cases}$$

We say that almost all graphs (or almost all graphs with $m(n)$ edges) have the property V if

$$\lim_{n \to \infty} f(\mathbf{G}_n) = 1 \quad (\text{or } \lim_{n \to \infty} f(\mathbf{G}_{nm}) = 1, \quad \text{respectively}). \quad ●$$

When performing an analysis of the average behaviour of algorithms it is desirable to know the properties common to almost all graphs. The basic papers in this direction are those of P Erdös and A Renyi [93, 94, 95], and the book [13]. Some of the results of these papers are very interesting and we are going to mention them here, but we do not present all proofs because they require a good knowledge of probability theory, and are often very difficult.

The first of the results concerns an upper bound on the size of a full subgraph.

Theorem 10.4.4. *Let $\varepsilon > 0$ be a real number. Then the following holds for almost all graphs G: if the graph G has n vertices, then it does not contain a clique of more than $2(1 + \varepsilon) \log n$ elements.*

Proof. Suppose X is a subset of the set $1, ..., n$. Let k denote the number of elements of X.

There exist exactly $\exp\binom{k}{2}$ graphs with the set X of vertices, and only one of them is complete. Consequently, the ratio of the number of graphs in which X is a clique to all graphs on the set $1, ..., n$ is equal to $\exp\left(-\binom{k}{2}\right)$; in other words, the number of such graphs is $\exp\left(\binom{n}{2} - \binom{k}{2}\right)$.

For a given k there exist $\binom{n}{k}$ subsets of the set $\{1, ..., n\}$ with k elements. The number of graphs in which a clique of k elements exists is thus at most $\binom{n}{k} \exp\left(\binom{n}{2} - \binom{k}{2}\right)$. The ratio of the last number and the number $\exp\binom{n}{2}$ of all graphs in the set $\{1, ..., n\}$ is, consequently, $\binom{n}{k} \exp\left(-\binom{k}{2}\right)$.

Thus, for $k > 2(1 + \varepsilon) \log n$ we have

$$\binom{n}{k} \exp\left(-\binom{k}{2}\right) \leq n^k \exp\left(-\binom{k}{2}\right) = \exp\left(k \log n - k \frac{k-1}{2}\right)$$

$$= \exp k \left(\log n - \frac{k-1}{2}\right) \leq \exp k \left(\frac{1}{2} - \varepsilon \log n\right)$$

and the limit of the last expression is 0 for $n \to \infty$. ●

Graphs with a large clique are, consequently, a rare exception. The size of cliques in graphs was studied in more detail by D W Matula [202] and B Bollobás and P Erdös [63] who proved that almost all graphs have, for a given set of vertices, one or two possible values of the size of a maximum clique, and hence, with a few exceptions, the number $\gamma(G)$ is almost completely determined by the number of vertices of the graph G, regardless of the position of its edges.

Theorem 10.4.5 [63]. *There exists a function $d = d(n)$ such that for each $\varepsilon > 0$ and for almost all graphs G the following holds: if G has n vertices, then*

$$d(n) - 1 - \varepsilon < \gamma(G) < d(n) + \varepsilon. \quad ●$$

It can be proved that the above function d has the following property:

$$d(n) = 2 \log n - 2 \log \log n + 2 \log (n/2) + o(1).$$

In precisely the same way as for theorem 10.4.4, it is possible to prove the following statement.

Theorem 10.4.6. *Let $\varepsilon > 0$ be a real number. Then the following holds for almost all graphs G: if the graph G has n vertices, then it does not have an independent set of more than $2(1 + \varepsilon) \log n$ vertices.* ●

As a consequence, we can easily determine the following estimate for the chromatic number of almost all graphs.

Theorem 10.4.7. *Let $\varepsilon > 0$ be a real number. Almost all graphs G have the following property: if G has n vertices, then*

$$\chi(G) > \frac{n}{2(1 + \varepsilon) \log n}.$$

Proof. If a graph of n vertices is coloured by k colours, then there exists a set of n/k vertices coloured by one colour. By the definition of graph colouring, the set must be independent, and we can apply theorem 10.4.6. ●

Several results are known for graphs with a small number of edges.

Theorem 10.4.8. *There exist constants* $c_1, c_2, ..., c_5$ *such that for* $m(n) = cn$, *where* c *is a constant, almost all graphs of* n *vertices and* $m(n)$ *edges have the size of the largest component bounded as follows:*

$$\begin{array}{lll} \text{between } c_1 \log n & \text{and } c_2 \log n & \text{if } 0 < c < \tfrac{1}{2}, \\ \text{between } c_3 n^{2/3} & \text{and } c_4 n^{2/3} & \text{if } c = \tfrac{1}{2}, \\ \text{between } c_5 n & \text{and } n & \text{if } c > \tfrac{1}{2}. \end{array}$$ ●

Theorem 10.4.9. *There exist constants* c_1 *and* c_2 *with* $0 < c_1 < c_2$ *such that whenever* $m(n) < c_1 n \log n$ *for all* n, *then almost all graphs of* n *vertices and* $m(n)$ *edges contain a vertex of degree 0 or 1, and hence, do not contain a hamiltonian circuit, whereas whenever* $m(n) > c_2 n \log n$ *for all* n, *then almost all graphs of* n *vertices and* $m(n)$ *edges contain a hamiltonian circuit.* ●

The last two theorems can be interpreted as follows: suppose that a certain n-point set is given, representing a graph without edges. We shall successively connect randomly chosen pairs of elements of that set by edges, and we shall investigate the properties of the graph constructed in this way. The basic piece of information which follows from theorems 10.4.8 and 10.4.9, as well as a number of similar results, is that quantitative and qualitative changes take place in most of the graphs at approximately the same instant, i.e. after approximately the same number of inserted edges, regardless of the way in which the edges were inserted. For example, after inserting approximately n edges the graph is no longer formed by isolated vertices and isolated edges, at the instant of inserting about $n/2$ edges a rapid growth of one of the components of the graph takes place (from approximately $\log n$ vertices to a number of vertices comparable with the number n of all vertices of the graph), a hamiltonian circuit will almost always appear when the number of edges of the graph is about $n \log n$, etc. Those theoretical results have also been verified by a computer, see, for example, [171], and they show that the properties of graphs, and consequently the usual behaviour of algorithms for processing them, depend essentially on the density of the graph.

10.5 Algorithms and random graphs

In Chapter 7 we tried to explain why for a very large number of problems concerning graphs no algorithms are known which would be capable of finding a solution rapidly even in the worst case. In Chapter 9 some of the approximate fast algorithms for solving difficult problems concerning graphs were presented. For most such algorithms, however, it is possible to show that in the worst case the results obtained are far removed from the optimum results, or no result is obtained at all. On the other hand, from the practical point of view it is usually sufficient that the algorithm be fast and sufficiently precise in a majority of cases. In the present section we thus return to the considerations mentioned in Chapter 9, and we shall investigate the behaviour of the algorithms in the average case. Most proofs will only be indicated or will be omitted altogether, since we do not assume that the reader is familiar with the deeper parts of probability theory.

We will first have a look at the algorithms for an approximate determination of the chromatic number of a graph. We are going to show that even the simple methods lead to rather good results on average.

Theorem 10.5.1. *Given a real number $\varepsilon > 0$, then for almost all graphs G the algorithm of 9.1.1(i) will find a colouring of G by at most $2(1 + \varepsilon)\chi(G)$ colours.*

Sketch of the proof. Let us have a look at the number of vertices which will be coloured by the first colour. The first vertex coloured by it is connected by an edge with $(n - 1)/2$ vertices on average. Considering the first i vertices coloured by the first colour, there exist around $n/2^i$ vertices which are not connected with any of those, and hence, which can be coloured by the first colour. Thus we may use the first colour for colouring vertices as long as for the number i of vertices thus coloured we have $n/2^i > 1$, and consequently, on average, we shall have around $\log n$ vertices coloured by that colour. Disregarding some technical difficulties we can say the same about the other colours, and hence, we will use $n/\log n$ colours on average. The statement above is proved by comparing this result with the estimate proof in theorem 10.4.7. ●

The same result can be proved for all of the known algorithms for colouring of graphs. On the other hand, it is not known whether there exists some approximate algorithm which would colour graphs in a polynomial-bounded running time and which would use for almost all graphs at most $2(1 - \varepsilon)\chi(G)$, or even at most $(1 + \varepsilon)\chi(G)$ colours.

We now turn to the search for a hamiltonian path. L Pósa has proved the following result which we give without proof.

Theorem 10.5.2. *There exists a positive constant c such that in almost all graphs of n vertices and cn* log *n edges, algorithm 9.2.1 will find a hamiltonian circuit.* ●

By comparing theorems 10.5.2 and 10.4.8, we see that the algorithm mentioned will find a hamiltonian circuit in almost all cases in which we can deduce its existence from the number of vertices and edges.

The travelling salesman problem is investigated in the paper [169] for the following probability distribution: n points are randomly distributed in a square region of the plane, and their distance is determined by the usual euclidean metric. An algorithm is presented which for each real number $\varepsilon > 0$ finds an approximate solution of the travelling salesman problem in such a way that it is probable that the precision of the solution obtained is better than $1 + \varepsilon$. For a fixed ε the algorithm requires running time $O(n \log n)$, but its speed depends heavily on the required precision. This result is based on the paper [55], where it is proved that the mean value of the shortest path of the travelling salesman is, in the case considered, proportionate to $\sqrt{n}$.

The average behaviour of algorithm 9.4.2 for searching for isomorphism of graphs is also known. The question of how fine is the division of vertices it constructs is essential for the length of computation. The following theorem is proved in the paper [42].

Theorem 10.5.3. *Algorithm 9.4.2 will divide the vertices of almost all graphs in such a way that all the classes of the resulting quasi-ordering are singleton sets.*

Sketch of the proof. The average degree of a vertex v of a graph of n vertices is a random variable of mean value $(n - 1)/2$ and variance of order $\sqrt{n}$. Since n is the number of all vertices, it is very probable that all the numbers between $(n - 1)/2 - \sqrt{n}$ and $(n - 1)/2 + \sqrt{n}$ are actual degrees of vertices of the graph, and hence the first division of the vertices of the graph according to degree will have at least $2n$ classes in almost all graphs. The probability that two distinct vertices of the graph have the same relationship to all these classes is very small (of order $n \exp(-n)$) and hence the second iteration of the division of vertices will already yield a complete differentiation of all vertices for almost all graphs. ●

The validity of theorem 10.5.3 has also been verified by practical experiments on a computer: for each $n = 35, 50, 70, 100, 140$ and 200, 200 experiments were performed and in each of them the algorithm returned a complete differentiation of the vertices after the second iteration.

Only for small values $n \leq 30$ did it happen that more than two iterations had to be performed in order to find a complete differentiation.

Further results concerning the average behaviour of algorithms processing graphs are also known; for example, finding the transitive closure of a graph in mean running time which is linear [233], finding the shortest path connecting each pair of vertices in mean running time $O(n^2 \log n \log n)$ [62], determining the components of a graph in linear mean running time [175], etc. In all of those cases it turns out that, from the point of view of average behaviour, even the simplest algorithms are satisfactory for solution of NP-complete problems and other types of difficult problems. The advantage of such algorithms is usually that they are fast. The overall picture of solvability of problems of graph theory is thus much more optimistic than could be deduced from Chapter 7. It is, however, necessary to recall that the results of the probabilistic analysis are usually based on the hypothesis that the frequency of occurrence is the same for all graphs of the given number of vertices, or the given number of vertices and edges. By choosing a different way of assigning weights to graphs, the result of the analysis of the average behaviour can change completely.

For example, an analysis of the following experiment has been presented in [184]: we assign to the elements of the set $\{1, ..., n\}$ random numbers $q(1), ..., q(n)$ chosen as natural numbers of value not exceeding a given number k. We then form a graph on the set $\{1, ..., n\}$ in such a way that if $q(i) \neq q(j)$, an edge connecting i and j is created with probability $\frac{1}{2}$, and if $q(i) = q(j)$ then no edge is created. The resulting graph is then very similar to the random graph G_n, but it is ensured that q is a colouring by at most k colours. The chromatic number can be prescribed, by a choice of k, to be substantially smaller than the mean value for G_n which is of order $n/\log n$. We are interested to know whether the algorithm under study will discover this difference.

It turns out that, after deleting an arbitrarily chosen vertex x and all of its neighbours from the graph described above, the mean values of the degrees of vertices y such that $q(x) = q(y)$ will be changed in a different way from the mean values of the degrees of vertices y with $q(x) \neq q(y)$. Some algorithms, e.g. 9.1.1(i), do not notice that difference, and colour the graph in the same way as the random graph G_n. Other, more refined, procedures, such as 9.1.3(iii), are able to discern the difference which is of order n/k, and will make use of it in order to find either directly the colouring q or at least a sufficiently near approximation. However, this capability of the mentioned algorithms takes place only when the above difference substantially exceeds the mean value of the variance of degrees of vertices from their averages. Since these variances are of order $\sqrt{n}$, we see that for

$k \ll \sqrt{n}$ algorithm 9.1.3(iii) should give excellent results: in the region $k \sim \sqrt{n}$ its behaviour should start changing quickly, and for $\sqrt{n} \ll k$ it should not differ much from the other algorithms. This has been confirmed by practical experiments. The following table shows the ratio of the number of colours used by the above algorithms to colour a graph to the chromatic number of the graph. This has been done for $n = 100$ and some values of k, averaging 15 experiments made for each k:

k	2	3	4	5	6	7	8	9	10
9.1.1(i)	1.30	2.51	2.85	2.70	2.63	2.33	2.19	2.00	1.89
9.1.3(iii)	1.00	1.02	1.03	1.17	1.20	1.36	1.50	1.65	1.63

10.6 Primality

G Miller has proved the following theorem in [209].

Theorem 10.6.1. *For arbitrary natural numbers n and k with $k < n$ denote by $\mathbf{T}(n, k)$ the following assertion:*
either $k^{n-1} \not\equiv 1 \bmod n$,
or there exists i such that the number $m = (n-1)/2^i$ is an integer, and the greatest common divisor of the numbers $k^m - 1$ and n is larger than 1 but smaller than n.
Then the number n is a prime if, and only if, there does not exist a number k such that $\mathbf{T}(n, k)$ is valid. ●

An application of the above theorem for a direct test of primality of the number n would clearly be unsuitable since it would be necessary, when n is indeed a prime, to consider all the numbers k from 1 to $n - 1$. It is, however, also proved in the above paper that assuming the validity of the generalised Riemann hypothesis, for each composite number n there exists k such that $\mathbf{T}(n, k)$ holds and $k \leq c(\log n)^2$ where c is a constant depending on the way in which the generalised Riemann hypothesis is valid. (We do not present the exact formulation of the hypothesis mentioned, since this would not be of any value for a reader with no experience in number theory; the reader may consult the paper mentioned.) In that case it would clearly be sufficient to test a small number of ks only, and hence, the resulting algorithm would work in running time bounded by a polynomial which depends on the number of binary digits of the given number n. This would deprive us of the last important example of a problem which lies in the class $NP \cap coNP$, but of which we do not know whether it lies in P or not.

In the present section we are interested in another algorithm which can be derived from the test of validity of the assertions $\mathbf{T}(n, k)$, and which does not require the assumption that the generalised Riemann hypothesis is valid. It can be proved that the following set:

$$\{k \mid k < n \quad \text{and} \quad \mathbf{T}(n, k) \text{ is valid}\}$$

has at least $(n - 1)/2$ elements for each composite n. Consequently, the probability that a random choice of k will be 'correct' is quite high. This has been used by M Rabin [222] for the following algorithm.

Algorithm 10.6.2. Primality test of natural numbers

Choose m random and mutually independent numbers $k_1, ..., k_m$ smaller than the number n being tested. If the assertion $\mathbf{T}(n, k_i)$ is valid for some i, then n is a composite and if $\mathbf{T}(n, k_i)$ does not hold for any $i = 1, ..., m$, then n is probably a prime.   ●

Let us estimate the probability of a wrong result which means that n is a composite number, but we have not managed to find a k such that $\mathbf{T}(n, k)$ is valid. The probability that we will miss once is, according to the above result, at most $1/2$, and hence, in the case of independent choices of m numbers k_i, the probability of failure is at most 2^{-m}, which is very small.

The above paper also presents some interesting applications: in several minutes all Mersenne primes $2^p - 1$ for $p \leq 500$ have been found without a single mistake, and the largest prime smaller than 2^{400} was shown to be $2^{400} - 593$ with probability of error smaller than 0.1 per cent.

Bibliography

1. Journals and proceedings

The theory of computational complexity of problems of graph theory and operational analysis has been developing rapidly, and the most important source of information is thus represented by journals and proceedings of conferences.

Let us mention some of the annual conferences:

Annual ACM Symposium on the Theory of Computing (STOC)
Annual Symposium on the Foundations of Computer Science (FOCS)
International Colloquium on Automata, Languages and Programming (ICALP)
Mathematical Foundation of Computer Science (MFCS)

The first two of these conferences are held in the USA, and individual proceedings are published for them, whereas the proceedings of the conferences ICALP, held in Western Europe, and MFCS, held alternately in Czechoslovakia and Poland, are published in the series *Lecture Notes of Computer Science* by the publisher Springer Verlag (Berlin, Heidelberg, New York).

The conference Fundamentals of Computation Theory (FCT) is biennial, and it has taken place in Poland, the German Democratic Republic and Hungary.

Concerning journals, we mention only those that are nearest to the aim of our book; others can be found in the list of journal papers below.

Communications of the Association for Computing Machinery (CACM)
Computing
IEEE Transactions. Computers

Information Processing Letters
Journal of the Association for Computing Machinery (JACM)
Journal of Computer and System Science (JCSS)
Kibernetika
Management Science
Mathematical Programming
Networks
Operations Research
SIAM Journal on Computing
Soviet Doklady (*of the Academy of Sciences*)
Theoretical Computer Science (TCS)

A further important source is the above-mentioned series *Lecture Notes in Computer Science*, published by Springer.

2. Monographs

[1] Aho A V, Hopcroft J E and Ullman J D 1974 *The Design and Analysis of Computer Algorithms* (Reading, MA: Addison-Wesley)

[2] Bellman R 1957 *Dynamic Programming* (Princeton, NJ: Princeton University Press)

[3] Berge C 1958 *Théorie des Graphes et ses Applications* (Paris: Gauthier-Villars)

[4] Berge C 1973 *Graphs and Hypergraphs* (Amsterdam: North-Holland)

[5] Berge C and Ghouila-Houri A 1965 *Programming, Games and Transportation Networks* (New York: John Wiley)

[6] Berztiss A T 1971 *Data Structures – Theory and Practice* (New York: Academic)

[7] Bollobás B 1979 *Graph Theory – An Introductory Course* (New York: Springer Verlag)

[8] Bondy J A and Murty U S R 1976 *Graph Theory with Applications* (New York: Elsevier)

[9] Busacker R G and Saaty T L 1968 *Finite Graphs and Networks* (New York: McGraw-Hill)

[10] Christofides N 1975 *Graph Theory – An Algorithmic Approach* (New York: Academic)

[11] Conway R W, Maxwell W L and Miller L W 1967 *Theory of Scheduling* (Reading, MA: Addison-Wesley)

[12] Deo N 1974 *Graph Theory with Applications to Engineering and Computer Science* (Englewood Cliffs, NJ: Prentice Hall)

[13] Erdös P and Spencer J 1974 *Probabilistic Methods in Combinatorics* (Budapest –Amsterdam: Akademiai Kiadó–North Holland)

[14] Even S 1973 *Algorithmic Combinatorics* (New York: Macmillan)

[15] Ford L R Jr and Fulkerson D R 1962 *Flows in Networks* (Princeton, NJ: Princeton University Press)

[16] Frank H and Frisch I 1972 *Connection, Transportation and Flow Networks* (Reading, MA: Addison-Wesley)

[17] Garey M R and Johnson D S 1979 *Computers and Intractability – A Guide to the Theory of NP-Completeness* (San Francisco: Freeman)

[18] Harary F 1969 *Graph Theory* (Reading, MA: Addison–Wesley)

[19] Hopcroft J E and Ullman J D 1969 *Formal Languages and their Relation to Automata* (Reading, MA: Addison-Wesley)

[20] Hu T C 1969 *Integer Programming and Network Flows* (Reading, MA: Addison-Wesley)

[21] Hu T C 1982 *Combinatorial Algorithms* (Reading, MA: Addison-Wesley)

[22] Knuth D E 1968 *The Art of Computer Programming, vol.* 1: *Fundamental Algorithms* (Reading, MA: Addison-Wesley)

[23] Knuth D E 1973 *The Art of Computer Programming, vol.* 3: *Sorting and Searching* (Reading, MA: Addison-Wesley)

[24] Lawler E L 1976 *Combinatorial Optimalization – Networks and Matroids* (New York: Holt, Reinhart and Winston)

[25] Lovasz L 1979 *Combinatorial Problems and Exercises* (Budapest–Amsterdam: Akademiai Kiadó–North Holland)

[26] Markov A A 1954 *Teorija algoritmov* (Trudy matematitcheskogo instituta imeni V A Steklova 42, Moscow, Izd. Acad. Nauk)

[27] Nečas J 1978 *Graphs and their Applications* (in Czech) (Prague: SNTL-Publishers of Technical Literature)

[28] Nešetřil J 1979 *Graph Theory* (in Czech) (Prague: SNTL-Publishers of Technical Literature)

[29] Nijenhuis A and Wilf H S 1975 *Combinatorial Algorithms* (New York: Academic)

[30] Ore O 1962 *Theory of Graphs* (New York: American Mathematical Society, College Publishing)

[31] Papadimitriou Ch and Steiglitz K 1982 *Combinatorial Optimization, Algorithms and Complexity* (Englewood Cliffs, NJ: Prentice Hall)

[32] Paul W 1978 *Komplexitätstheorie* (Stuttgart: Teubner Verlag)

[33] Rajlich V 1979 *Introduction to the Theory of Computers* (in Czech) (Prague: SNTL-Publishers of Technical Literature)

[34] Reinhold E M, Nievergelt J and Deo N 1977 *Combinatorial Algorithms, Theory and Practice* (Englewood Cliffs, NJ: Prentice Hall)

[35] Rodgers H Jr 1967 *Theory of Recursive Functions and Effective Computability* (New York: McGraw-Hill)

[36] Savage J 1976 *The Complexity of Computing* (New York: John Wiley)

[37] Sedláček J 1964 *Combinatorics in Theory and Practice* (in Czech) (2nd edn) (Prague: Publishing House of the Academy of Sciences)

[38] Standish T A 1980 *Data Structure Techniques* (Reading, MA: Addison-Wesley)

[39] Tarjan R E 1983 *Data Structures and Network Algorithms* (Philadelphia, PA: SIAM)

[40] Wirth N E 1975 *Algorithms + Data Structures = Programs* (Englewood Cliffs, NJ: Prentice Hall)

[41] Zykov A A 1969 *Theory of Finite Graphs* (in Russian) (Novosibirsk: Nauka)

3. Papers and technical reports

[42] Adelson-Velskij G M and Landis Yu M 1962 An algorithm of organization of information. (in Russian) *Sov. Dokl.* **146** 263-6

[43] Adleman L 1978 Two theorems on random polynomial time. *19th FOCS* 75–84

[44] Adleman L and Manders K 1977 Reducibility, randomness and intractability. *9th STOC* 151–63

[45] Angluin D and Valiant L G 1977 Fast probabilistic algorithms for hamiltonian circuits and matchings. *9th STOC* 30–4

[46] Appel K and Haken W 1976 Every planar map is four colorable. *Bull. Am. Math. Soc.* **82** 711–12

[47] Aris D 1983 A survey of heuristics for the weighted matching problem. *Networks* **13** 475–93

[48] Arlazarov V L, Dinic E A, Kronod M A and Faradzev I A 1970 Economical construction of the transitive closure of a graph. (in Russian) *Sov. Dokl.* **194** 487–8

[49] Auslander L and Parter S V 1961 On imbedding graphs into the plane. *J. Math. Mech.* **10** 517–23

[50] Babai L 1979 Monte-Carlo algorithms in graph isomorphism, testing. Preprint, University of Montreal

[51] Babai L, Grigorijev D Yu and Mount D 1982 Isomorphism testing for graphs with bounded eigenvalue multiplicities. *14th STOC* 310–24

[52] Babai L and Kučera L 1979 Canonical labelling of graphs in linear average time. *20th FOCS* 39–46

[53] Baker B S 1983 Approximation algorithms for NP-complete problems on planar graphs. *24th FOCS* 265–72

[54] Baker T, Gill J and Solovay R 1975 Relativization of the $P = ?NP$ question. *SIAM J. Comput.* **4** 431–42

[55] Bearwood J, Halton J H and Hammersley J M 1959 The shortest path through many points. *Proc. Camb. Phil. Soc.* **55** 299–327

[56] Bellman R 1958 On a routing problem. *Q. Appl. Mech.* **16** 87–90

[57] Bent S W, Sleator D D and Tarjan R E 1980 Biased 2–3 Trees. *21st FOCS* 248–54

[58] Bent S W, Sleator D D and Tarjan R E 1985 Biased search trees. *SIAM J. Comput.* **14** 545–68

[59] Bentley J L 1979 Decomposable searching problems. *Info. Proc. Lett.* **8** 244–51

[60] Berge C 1957 Two theorems in graph theory. *Proc. Natl. Acad. Sci. USA* **43** 842–4

[61] Berman L and Hartmanis J 1977 On isomorphism and density of NP and other complete sets. *SIAM J. Comput.* **6** 305–22

[62] Bloniarz P 1983 A shortest path algorithm with expected time $O(n^2 \log n \log^* n)$. *SIAM J. Comput.* **12** 588–600

[63] Bollobás B and Erdös P 1976 Cliques in random graphs. *Math. Proc. Camb. Phil. Soc.* **80** 419–27

[64] Booth K S and Colbourn C J 1977 Problems polynomially equivalent to graph isomorphism. *Tech. Rep. CS-77-D4* University of Waterloo, Dept. of Computer Science

[65] Borůvka O 1926 On a certain minimum problem. (in Czech) *Papers of the*

Moravian Science Society in Brno (*Práce moravské přirodovědecké společnosti v Brně*) **3** 37–58

[66] Brélaz D 1979 New methods to color the vertices of a graph. *CACM* **22** 251–6

[67] Brown J R 1972 Chromatic scheduling and the chromatic number problems. *Management Sci.* **19** 456–63

[68] Chandra A K and Stockmeyer L 1976 Alternation. *17th FOCS* 98–108

[69] Chatchiian L G 1979 Polynomial algorithm in linear programming. (in Russian) *Sov. Dokl.* **244** 1093–6

[70] Cheriton D and Tarjan R E 1976 Finding minimum spanning trees. *SIAM J. Comput.* **5** 724–42

[71] Choquet G 1938 Etude de certain réseaux de routes. *C. R. Acad. Sci., Paris* **206** 310–13

[72] Christofides N 1980 The travelling salesman problem – a survey. *Tech. Rep. Imperial College* (London)

[73] Chytil M 1977 Optimal decision-making. (in Czech) *Proc. SOFSEM'77*, (Bratislava: University Publishing House) 43–68

[74] Colbourn C J 1978 A bibliography of the graph isomorphism problem. *Tech. Rep.* 123/78 University of Toronto

[75] Cook S A 1971 The complexity of theorem proving procedures. *3rd STOC* 151–8

[76] Corneil D G and Gotlieb C C 1970 An efficient algorithm for graph isomorphism. *JACM* **17** 51–64

[77] Cosmadakis S S and Papadimitriou C H 1984 The traveling salesman problem with many visits to few cities. *SIAM J. Comput.* **13** 99–108

[78] DeMillo R and Lipton R 1980 The consistency of 'P = NP' and related problems with a fragment of number theory. *12th STOC* 45–57

[79] Demoucron G, Malgrange Y and Pertuiset R 1964 Graphes planaires. Reconnaissance et construction des representations planaires topologiques. *Revue Francaise de Recherche Operationelle* **8** 34–47

[80] Deo N 1976 Note on Hopcroft and Tarjan's planarity algorithm. *JACM* **23** 74–5

[81] Deroye L 1986 A note on the height of binary search trees. *JACM* **33** 489–98

[82] Dijkstra E W 1959 A note on two problems in connection with graphs. *Numer. Math.* **1** 269–71

[83] Dinic E A 1970 Algorithm of solution of the problem of maximum flow in a network with polynomial bound. (in Russian) *Sov. Dokl.* **194** 754–7

[84] Dinic E A 1973 On finding maximum flow in networks of a special type and some applications. (in Russian) In *Mathematical Questions of Control*, vol. 5, (Moscow: MGU Publishing House)

[85] Dreyfus S E 1969 An appraisal of some shortest path algorithms. *Oper. Res.* **17** 395–412

[86] Dyer M E and Frieze A M 1986 Fast solution of some random NP-hard problems. *27th FOCS* 331–6

[87] Edmonds J 1965 Paths, trees and flowers. *Can. J. Math* **17** 449–67

[88] Edmonds J 1965 Maximum matching and a polyhedron with 0, 1 vertices. *J. Res. NBS* **B69** 125–30

[89] Edmonds J 1965 The Chinese Postman problem. *Oper. Res.* **13** Suppl. 1, 373

[90] Edmonds J and Johnson E L 1970 Matching: A well solved class of integer linear programs. In *Combinatorial Structures and Their Applications* ed. R Guy (New York: Gordon and Breach) pp 89–91

[91] Edmonds J and Johnson E L 1973 Matching. Euler Tours and the Chinese Postman. *Math. Programming* **5** 88–124

[92] Edmonds J and Karp R M 1972 Theoretical improvements on algorithmic efficiency for network flow problem. *JACM* **19** 248–64

[93] Erdös P 1939, 1961 Graph theory and probability. *Can. J. Math.* I: **11** 34–8, II: **13** 346–52

[94] Erdös P and Renyi A 1959 On random graphs. *Publ. Math. Debrecen* **6** 290–7

[95] Erdös P and Renyi A 1960 On evolution of random graphs. *Math. Kutató Int. Közl.* **5** 17–60

[96] Esfahanian A and Hakimi S 1984 On computing the connectivities of graphs and diagraphs. *Networks* **14** 353–66

[97] Even S and Hopcroft J E 1975 An algorithm for determining whether the connectivity of a graph is at least k. *SIAM J. Comput.* **4** 396

[98] Even S and Kariv O 1975 An $O(n^{5/2})$ algorithm for maximum matchings in general graphs. *16th FOCS* 100–12

[99] Even S and Tarjan R E 1975 Network flow and testing graph connectivity. *SIAM J. Comput.* **4** 507–18

[100] Even S and Tarjan R E 1976 A combinatorial problem which is complete in polynomial space. *JACM* **23** 710–19

[101] Filotti I S and Mayer J N 1980 A polynomial-time algorithm for determining the isomorphism of graphs of fixed genus. *12th STOC* 236–43

[102] Filotti I S, Miller G L and Reif J H 1979 On determining the genus of a graph in $O(v^{0(g)})$ steps. *11th STOC* 27–37

[103] Fisher M J and Meyer A R 1971 Boolean matrix multiplication and transitive closure. *12th Symp. on Switching and Automata Theory (now FOCS)* 129–31

[104] Flajolet P and Odlyzkol A 1982 The average height of binary trees and other simple trees. *JCSS* **25** 171–213

[105] Floyd R W 1962 Algorithm 97: shortest path. *CACM* **5** 345

[106] Ford L R and Fulkerson D R 1956 Maximal flow through a network. *Can. J. Math.* **8** 399–404

[107] Fraenkel A S, Barey M R, Johnson D S, Schaefer T and Yesha Y 1978 The complexity of checkers on a $N \times N$ board. Preliminary Report. *19th FOCS* 55–64

[108] Fredericson G 1987 A new approach to all pairs shortest paths in planar graphs. *19th STOC* 19–28

[109] Fredericson G 1983 Shortest path problem in planar graphs. *24th FOCS* 242–7

[110] Fredman L 1976 New bounds on the complexity of the shortest path problem. *SIAM J. Comput.* **5** 87–9

[111] Fredman M and Tarjan R E 1984 Fibonacci heaps and their use in improved network optimization algorithms. *25th FOCS* 338 46

[112] Frieze A M 1983 An extension of Christofides heuristic to the *k*-person travelling salesman problem. *Discr. Appl. Math.* **6** 79–93

[113] Gabow H N 1976 An efficient implementation of Edmonds' algorithm for maximum matching on graphs. *JACM* **23** 221–34

[114] Gabow H 1985 Scaling algorithms for network problems. *JCSS* **31** 149–68

[115] Gabow H 1985 A scaling algorithm for weighted matching on general graph. *26th FOCS* 90–100

[116] Gabow H, Galil Z and Spencer T 1984 Efficient implementation of graph algorithms using contractions. *25th FOCS* 347–57

[117] Galil Z 1978 A new algorithm for the maximal flow problem. *19th FOCS* 231–45

[118] Galil Z, Hoffman C M, Luks E M, Schnorr C P and Weber A 1982 An $O(n^3 \log n)$ deterministic and an $O(n^3)$ probabilistic isomorphism test for trivalent graphs. *23rd FOCS* 118–25

[119] Galil Z, Micali S and Gabow H 1986 Priority queues with variable priority and $O(FV \log V)$ algorithm for finding a maximal weighted matching in general graphs. *SIAM J. Comput.* **15** 120–30

[120] Galil Z and Naamad A 1980 An $O(EV \log^2 V)$ algorithm for the maximal flow problem. *JCSS* **21** 203–17

[121] Galil Z and Tardos E 1986 An $O(n^2(m + n \log n))$ min cost flow algorithm. *27th FOCS* 1–9

[122] Garey M R, Graham R L and Johnson D S 1976 Some *NP*-complete geometric problems. *8th STOC* 10–22

[123] Garey M R and Johnson D S 1974 Complexity results for multiprocessor scheduling under resource constraints. *SIAM J. Comput.* **3** 397–411

[124] Garey M R and Johnson D S 1976 Approximation algorithms for combinatorial problems. In *An Annotated Bibliography, in Algorithms and Complexity* ed. J F Traub (New York: Academic) pp 41–52

[125] Garey M R and Johnson D S 1976 The complexity of near optimal graph colouring. *JACM* **23** 43–69

[126] Garey M R and Johnson D S 1977 Two processor scheduling with start times and deadlines. *SIAM J. Comput.* **6** 416–26

[127] Garey M R and Johnson D S 1977 The rectilinear Steiner tree problem is *NP*-complete. *SIAM J. Appl. Math.* **32** 826–34

[128] Garey M R and Johnson D S 1978 Motivations, examples and implications. *JACM* **25** 499–508

[129] Garey M R, Johnson D S and Stockmeyer L 1976 Some simplified *NP*-complete graph problems. *TCS* **1** 237–67

[130] Garey M R, Johnson D S and Tarjan R E 1976 The planar hamiltonian circuit problem is *NP*-complete. *SIAM J. Comput.* **5** 704–14

[131] Gilbert J R, Lengauer T and Tarjan R E 1969 The pebbling problem is complete in polynomial space. 11*th STOC* 237–48

[132] Gill J 1977 Computational complexity of probabilistic Turing machines. *SIAM J. Comput.* **6** 675–95

[133] Goldberg A and Tarjan R E 1985 Self-adjusting binary search trees. *JACM* **32** 652–86

[134] Goldberg A and Tarjan R E 1987 Solving minimum cost flow problem by successive approximation. 19*th STOC* 7–18

[135] Goldstein A J 1963 An efficient and constructive algorithm for testing whether a graph can be embedded into a plane. *Graph and Combinatorics Conf.* Princeton University, p 2

[136] Grimmet G R and McDiarmid C J H 1975 On colouring random graphs. *Math. Proc. Camb. Phil. Soc.* **77** 313–24

[137] Grötschel M, Lovász L and Schrijver S 1980 The ellipsoid method and its consequences in combinatorial optimization. *Tech. Rep. Univ. Bonn*

[138] Grötschel M and Padberg M W 1979 On the symmetric travelling salesman problem. *Math. Programming* **16** 265–302

[139] Gruska J 1975 Complexity of concrete algorithms I: basic principles. (in Slovak) *Lecture Notes* 2 (Bratislava: University Publishing House)

[140] Gruska J 1976 Complexity of concrete algorithms II: sorting algorithms. (in Slovak) *Lecture Notes* 13 (Bratislava: University Publishing House)

[141] Gruska J, Wiederman J and Černý A 1978 Types and structures of data. (in Slovak) *Proc. SOFSEM'78* (Bratislava: University Publishing House)

[142] Gusfield D, Martel Ch and Fernandez-Basa D 1987 Fast algorithms for bipartite network flow. *SIAM J. Comput.* **14** 237–51

[143] Hájek P 1979 Arithmetical hierarchy and complexity of computation. *TCS* **8** 227–37

[144] Hamacher H 1979 Numerical investigations of the maximal flow algorithm of Karzanov. *Computing* **22** 17–30

[145] Hartmanis J and Hopcroft J E 1976 Independence results in computer science. *SIGACT News* **8** 13–24

[146] Hassin R 1981 Maximal flow in (s, t)-planar network. *IPL* **107**

[147] Hassin R and Johnson D B 1985 An $O(n \log^2 n)$ algorithm for maximum flow undirect planar networks. *SIAM J. Comput.* **14** 612–24

[148] Held M and Karp R M 1970, 1971 The travelling salesman problem and minimum spanning trees. Part I *Oper. Res.* **18** 1138–62. Part II *Math. Programming* **1** 6–25

[149] Hoare C A R 1962 Quicksort *Comput. J.* **5** 10–15

[150] Hochbaum D S and Shoums D B 1987 A unified approach to approximation algorithms for bottleneck problems. *JACM* **33** 533–50

[151] Hopcroft J E and Karp R M 1973 An $n^{5/2}$ algorithm for maximum matching in bipartite graphs. *SIAM J. Comput.* **2** 255–31

[152] Hopcroft J E, Paul W and Valiant L 1977 On time versus space. *JACM* **24** 332–7

[153] Hopcroft J E and Tarjan R E 1973 Dividing a graph into triconnected components. *SIAM J. Comput.* **2** 135–58

[154] Hopcroft J E and Tarjan R E 1973 Efficient algorithm for graph manipulation. *CACM* **16** 372–8

[155] Hopcroft J E and Tarjan R E 1974 Efficient planarity testing. *JACM* **21** 549–68

[156] Hopcroft J E and Tarjan R E 1973 A $V \log V$ algorithm for isomorphism of triconnected planar graphs. *JCSS* **7** 323–31

[157] Hopcroft J E and Wong J K 1974 Linear time algorithm for isomorphism of planar graphs. *6th STOC* 172–84

[158] Ibarra O H and Kim C E 1975 Fast approximation algorithms for the knapsack and sum of subset problems. *JACM* **22** 463–8

[159] Itai A and Shiloach Y 1979 Maximal flow in planar networks. *SIAM J. Comput.* **8** 135–50

[160] Jarník V 1930 On a certain minimum problem. (in Czech) *Brno, Papers of the Moravian Science Society* **4** 57–63

[161] Johnson D B 1977 Efficient algorithms for shortest paths in sparse networks. *JACM* **24** 1–13

[162] Johnson D S *NP*-completeness column *J. Algorithms*

[163] Johnson D B and Venkatesan S M 1983 Partition of planar flow networks. *24th FOCS* 259–64

[164] Jones D 1975 Space-bounded reductions among combinatorial problems. *JCSS* **11** 68–85

[165] Joseph D and Young P 1980 Independence results in computer science? *JCSS 12th STOC* 58–69

[166] Kameda T and Munro I 1974 An $O(VE)$ algorithm for maximum matching of graphs. *Computing* **12** 91–8

[167] Kanevsky A and Ramachandran V 1987 Improved algorithms for graph four-connectivity *28th FOCS* 252–9

[168] Karp R M 1972 Reducibility among combinatorial problems. In *Complexity of Computer Computations* ed. E W Miler and J W Thatcher (New York: Plenum Press) pp 85–104

[169] Karp R M 1971 Probabilistic analysis of partitioning algorithms for the travelling salesman problem in the plane. *Tech. Rep. UCB/ERL M77/3¹* (University of California, Berkeley)

[170] Karp R M 1975 On the computational complexity of combinatorial problems. *Networks* **5** 45–68

[171] Karp R M 1976 The probabilistic analysis of some combinatorial search algorithms. In *Algorithms and Complexity* ed. J F Traub (New York: Academic)

[172] Karp R M 1979 A patching algorithm for the nonsymmetric travelling salesman problem. *SIAM J. Comput.* **8** 501–73

[173] Karp R M and Lipton R 1980 Some connections between non-uniform and uniform complexity classes. *12th STOC* 302–9

[174] Karp R M and Sipser M 1981 Maximum matching in sparse random graphs. *22nd FOCS* 364–75

[175] Karp R M and Tarjan R E 1980 Linear expected time algorithms for connectivity problems. *12th STOC* 368–77

[176] Karzanov A V 1974 Finding maximal flow in a network by means of pre-flows. (in Russian) *Sov. Dokl.* **215** 49–52

[177] Klee V and Minty G J 1972 How good is the simplex algorithm? In *Inequalities III* ed. O Shisha (New York: Academic) pp 159–75

[178] Knuth D E 1971 Optimal binary search trees *Acta Informatica* **1** 14–25

[179] Knuth D E and Jonassen A T 1978 A trivial algorithm whose analysis isn't. *JCSS* **16** 301–22

[180] Komlos J 1984 Linear verification for spanning trees. *25th FOCS* 201–6

[181] Korshunov A D 1976 Solution of the problem of Erdös and Renyi. (in Russian) *Sov. Dokl.* **228** 529–32

[182] Krishnamoorthy M S and Deo N 1979 Node deletion *NP*-complete problems. *Networks* **9** 189–94

[183] Kruskal J B Jr 1957 On the shortest spanning of a graph and the travelling salesman problem. *Proc. Am. Math. Soc.* **7** 48–50

[184] Kučera L 1977 Expected behaviour of graph coloring algorithms. *FCT, Lecture Notes in Computer Science* **56** 447–51

[185] Kučera L 1981 Maximal flow in planar networks. *MFCS'81, Lecture Notes in Computer Science* **118** 418–22

[186] Kuhn H W 1955 The Hungarian method for assignment problem. *Naval Res. Logistic Q.* **2** 83–97

[187] Levin L 1986 Average-case complete problems. *SIAM J. Comput.* **15** 285–6

[188] Lichtenstein D 1980 Isomorphism for graphs embeddable in the projective plane. *12th STOC* 218–24

[189] Lichtenstein D and Sipser M 1980 GO is PSPACE hard. *JACM* **27** 393–401

[190] Liks E M 1980 Isomorphism of graphs of bounded valence can be tested in polynomial time. *21st FOCS* 42–9

[191] Lin S and Kerninghan B W 1973 An effective heuristic algorithm for the travelling salesman problem. *Oper. Res.* **21** 498–516

[192] Lingas A 1978 A PSPACE-complete problem related to a pebble game. *5th ICALP, Lecture Notes in Computer Science* **62** 300–21

[193] Linial N, Lovasz L and Windgerson A 1986 A physical interpretation of graph connectivity, and its algorithmic application *27th FOCS* 39–48

[194] Lipton R and Tarjan R E 1979 A separator theorem for planar graphs *SIAM J. Appl. Math.* **36** 177–89

[195] Lubiw A 1981 Some *NP*-complete problems similar to graph isomorphism *SIAM J. Comput.* **10** 11–21

[196] Mahaney S R 1980 Sparse complete sets for NP: solution of a conjecture of Berman and Hartmanis. *21st FOCS* 54–68

[197] Mahaney S R 1981 On the number of P-isomorphism classes of NP-complete sets. *22nd FOCS* 271–8

[198] Malhotra V M, Pramodh Kumar M and Maheshwari S N 1978 An $O(V^3)$ algorithm for finding maximum flows in networks. *Info. Proc. Lett.* **7** 277–8

[199] Malhorn K and Tsakalidis A 1986 Self adjusting heaps. *SIAM J. Comput.* **15** 52–69

[200] Manders K and Adleman L 1978 NP-complete decision problems for binary quadratics. *JCSS* **16** 168–84

[201] Matimond H and Pittel B 1984 On the most probable shape of a binary search tree grown from a random permutation. *SIAM J. Algebr. Discr. Math.* **5** 69–81

[202] Matula D 1976 The largest clique size in a random graph. *Tech. Rep. Southern Methodist University, Texas*

[203] Matula D W 1987 Determining edge connectivity in $O(nm)$. 28*th FOCS* 249–51

[204] Matula D, Marble G and Isaacson J D 1972 Graph coloring algorithms. In *Graph Theory and Computations* ed. R C Read (New York: Academic)

[205] Mei-Ko Kwan 1962 Graphic programming using odd and even points. *Chinese Math.* **1** 273–7

[206] Melhorn K 1979 Dynamic binary search. *SIAM J. Comput.* **8** 175–98

[207] Micali S and Vazirani V V 1980 A $\sqrt{(V)}E$ algorithm for finding maximum matching in general graphs. 21*st FOCS* 17–27

[208] Miller A and Ramachandran V 1987 A new graph to connectivity algorithm and its parallelization. 19*th STOC* 345–54

[209] Miller G 1976 Riemann's hypothesis and test for primality. *JCSS* **13** 300–17

[210] Miller G L 1980 Isomorphism testing for graphs of bounded genus. 12*th STOC* 225–35

[211] Munro I 1971 Efficient determination of the transitive closure of a directed graph. *Info. Proc. Lett.* **1** 56–8

[212] Nievergelt J and Wong C K 1972 On binary search trees. *Proc. IVIP Congr. 1971* (Amsterdam: North Holland) pp 91–8

[213] Overmars M H and van Leeuwen J 1981 Two general methods for dynamizing decomposable search problems. *Computing* **26** 155–66

[214] Papadimitriou Ch 1976 The NP-completeness of the bandwidth minimization problem. *Computing* **16** 263–70

[215] Papadimitriou Ch 1977 The euclidean travelling salesman problem is *NP*-complete. *TCS* **4** 237–44

[216] Paul W and Tarjan R E 1978 Time-space trade-offs in a pebble game. *Acta Informatica* **10** 111–15

[217] Pierce A N 1975 Bibliography on algorithms for shortest path, shortest spanning tree and related circuit routing problems (1956 – 1974). *Networks* **5** 129–49

[218] Pittel B 1984 On growing random search trees. *J. Math. Anal. Appl.* **103** 461–80

[219] Pósa L 1976 Hamiltonian circuits in random graphs. *Discrete Math.* **14** 359–64

[220] Pratt V 1975 Every prime has a succinct certificate. *SIAM J. Comput.* **4** 214–20

[221] Prim R C 1957 Shortest connection networks and some generalizations. *Bell. Syst. Tech. J.* **36** 389–401

[222] Rabin M O 1976 Probabilistic algorithms. In *Algorithms and Complexity* ed. J F Traub (New York: Academic) pp 21–39

[223] Read R C and Corneil D G 1977 The graph isomorphism disease. *J. Graph Theory* **1** 339–63

[224] Reif J H 1981 Minimum S-T cut of a planar undirected network in $O(n \log^2 n)$ time. *ICLAP* 81, *Lecture Notes in Computer Science* **115** 56–67

[225] Reif J 1983 Minimum s-t cut a planar undirected network in $O(n \log^2 n)$ time. *SIAM J. Comput.* **12** 71–81

[226] Reingold E M and Tarjan R E 1981 On a greedy heuristic for complete matching. *SIAM J. Comput.* **10** 676–81

[227] Rivest R and Vuillemin J 1975 A generalization and proof of the Aandreaa-Rosenberg conjecture. *7th STOC* 6–11

[228] Rubin F 1975 An improved algorithm for testing the planarity of a graph. *IEEE Trans. Comput.* **C-24** 113–21

[229] Růžička P 1981 Time and memory in computational complexity. (in Slovak) *Information Systems (Informačné systémy)* **10** 139–58

[230] Savitch W J 1970 Relationship between nondeterministic and deterministic tape complexities. *JCSS* **4** 177–92

[231] Saxe J B and Bentley J L 1979 Transforming static data structures into dynamic structures. *20th FOCS* 148–168

[232] Schaefer T G 1978 Complexity of some perfect information games. *JCSS* **16** 185–225

[233] Schnorr C P 1978 An algorithm for transitive closure with linear expected time. *SIAM J. Comput.* **7** 127–33

[234] Sedgevick R and Vitter J 1984 Shortest paths in euclidean graphs. *25th FOCS* 417–42

[235] Sethi R 1975 Complete register allocation problems. *SIAM J. Comput.* **4** 226–48

[236] Sharir M 1987 On shortest path amidst convex polyhedra. *SIAM J. Comput.* **16** 561–72

[237] Shiloach Y 1978 An $O(nI \log^2 I)$ maximum flow algorithm. *Tech. Rep. STAN-CS-78-702* Stanford Univ.

[238] Sleator D D 1980 An $O(nm \log n)$ algorithm for maximum network flows. *Ph D Dissertation* Stanford University

[239] Sleator D D and Tarjan R E 1981 A data structure for dynamic trees. *13th STOC* 114–22

[240] Sleator D D and Tarjan R E 1983 A data structure for dynamic trees. *JCSS* **24** 362–91

[241] Sleator D D and Tarjan R E 1985 Self adjusting binary search trees. *JACM* **32** 652–86

[242] Sleator D D and Tarjan R E 1986 Self adjusting heaps. *SIAM J. Comput.* **15** 52–69

[243] Solovay R and Strassen V 1977 A fast Monte-Carlo test for primality. *SIAM J. Comput.* **6** 84–5

[244] Sozonov V Yu 1981 A logical approach to the problem '$P = NP$?'. *MFCS'80, Lecture Notes in Computer Science* **88** 562–75; Corrigendum: *MFCS'81, Lecture Notes in Computer Science* **118** 490

[245] Spira P 1973 A new algorithm for finding all shortest paths in a graph of positive edges in average time $O(n^2 \log^2 n)$. *SIAM J. Comput.* **2** 28–32

[246] Stockmeyer L J and Meyer A R 1973 Word problems requiring exponential time. *5th STOC* 1–9

[247] Strassen V 1969 Gaussian elimination is not optimal. *Numer. Math.* **13** 354–6

[248] Supowit K and Reingold E M 1983 Divide and conquer heuristics for minimum weighted Euclidean matching. *SIAM J. Comput.* **12** 118–43

[249] Supowit K J, Reingold E M and Plaisted D A 1983 The traveling salesman problem and minimum matching in the unit square. *SIAM J. Comput.* **12** 144–56

[250] Tarjan R E 1972 Depth-first search and linear graph algorithms. *SIAM J. Comput.* **1** 146–60

[251] Tarjan R E 1974 Finding dominators in directed graphs. *SIAM J. Comput.* **3** 62–89

[252] Tarjan R E 1974 Testing flow graph reducibility. *JCSS* **9** 335–65

[253] Tarjan R E 1975 Efficiency of a good but not linear set union algorithm. *JACM* **22** 215–25

[254] Tarjan R E 1977 Reference machines require non-linear time to maintain disjoint sets. *9th STOC* 18–29

[255] Tarjan R E 1978 Complexity of combinatorial algorithms. *SIAM Review* **20** 457–91

[256] Tarjan R E 1980 Recent developments in the complexity of combinatorial algorithms. *Proc. 5th IBM Symp. on Mathematical Foundations of Computer Science Japan* pp. 1–28

[257] Tarjan R E 1984 A simple version of Karzanov's blocking flow algorithm. *Oper. Res. Lett.* **2** 265–8

[258] Tcherkaskiy B V 1977 An algorithm for solving the problem of maximum flow in a network of complexity $O(V^2 \sqrt{E})$. (in Russian) *Mathematical Methods of Solving Economy Problems* **7** 117–25

[259] Ullman J D 1975 NP-complete scheduling problems. *JCSS* **10** 384–93

[260] Vaishnavi V K 1987 Weighted leaf AVL-trees. *SIAM J. Comput.* **16** 503–37

[261] Van Emde Boas P 1977 Preserving order in a forest in less than logarithmic time and linear space. *Info. Proc. Lett.* **6** 80–2

[262] Van Emde Boas P, Kaas R and Zijlstra E 1977 Design and implementation of an efficient priority queue. *Math. Syst. Theory* **10** 99–127

[263] Vuillemin J 1978 A data structure for manipulating priority queues. *CACM* **21** 309–15

[264] Wagner R 1976 A shortest path algorithm for edge sparse graphs. *JACM* **23** 50–7

[265] Weinberg L A 1966 Simple and efficient algorithm for determining isomorphism of planar triply connected graphs. *IEEE Trans. Circuit Theory* **CT-13** 142–8

[266] Welsh D J A and Powell M B 1967 An upper bound for the chromatic number of a graph and its application to large scale time tabling. *Problems Comput. J.* **10** 85–6

[267] Whitney H 1933 A set of topological invariants for graph. *Am. J. Math.* **55** 321–5

[268] Wiedermann J 1980 Probabilistic algorithms. (in Slovak) *Information Systems* (*Informačné systémy*) **9** 245–58

[269] Wiedermann J 1981 Preserving total order in constant expected time. In *MFCS'81, Lecture Notes in Computer Science* **118** 554–62

[270] Williams J W J 1964 Heapsort: algorithm 232. *CACM* **7** 347–8

[271] Williams M R 1970 The colouring of very large graphs. In *Combinatorial Structures and Their Applications* ed. R Cuy (New York: Gordon and Breach) pp. 477–8

[272] Yao A 1975 An $O(E \log \log V)$ algorithm for finding minimum spanning trees. *Info. Proc. Lett.* **4** 21–3

[273] Zadeh N 1973 A bad network problem for the simplex method and other minimum cost flow algorithms. *Math. Programming* **5** 255–66

[274] Zykov A A 1949 On some properties of linear complexes. (in Russian) *Matem. Sbornik* **24** 163–88

Index